Racial Identities, Genetic Ancestry, and
Health in South America

Racial Identities, Genetic Ancestry, and Health in South America

Argentina, Brazil, Colombia, and Uruguay

Edited by

**Sahra Gibbon,
Ricardo Ventura Santos, and
Mónica Sans**

For Nanneke
with thanks,
Sahra
January 2012

palgrave
macmillan

RACIAL IDENTITIES, GENETIC ANCESTRY, AND HEALTH IN SOUTH AMERICA
Copyright © Sahra Gibbon, Ricardo Ventura Santos, and Mónica Sans, 2011.

First published in 2011 by
PALGRAVE MACMILLAN®
in the United States—a division of St. Martin's Press LLC,
175 Fifth Avenue, New York, NY 10010.

Where this book is distributed in the UK, Europe and the rest of the world,
this is by Palgrave Macmillan, a division of Macmillan Publishers Limited,
registered in England, company number 785998, of Houndmills,
Basingstoke, Hampshire RG21 6XS.

Palgrave Macmillan is the global academic imprint of the above companies
and has companies and representatives throughout the world.

Palgrave® and Macmillan® are registered trademarks in the United States,
the United Kingdom, Europe and other countries.

ISBN: 978–0–230–11061–8

Library of Congress Cataloging-in-Publication Data

Racial identities, genetic ancestry, and health in South America :
Argentina, Brazil, Colombia, and Uruguay / edited by Sahra Gibbon,
Ricardo Ventura Santos, and Mónica Sans.
    p. cm.
Includes bibliographical references and index.
ISBN 978–0–230–11061–8
    1. Human population genetics—South America. 2. Genomics—South
America. 3. Anthropology—South America. 4. Medical genetics—South
America. 5. Sociobiology—South America. 6. Ethnology—South America.
I. Gibbon, Sahra. II. Santos, Ricardo Ventura, 1964– III. Sans, Mónica.

GN290.S63R33 2011
304'6098—dc23   2011020948

A catalogue record of the book is available from the British Library.

Design by Newgen Imaging Systems (P) Ltd., Chennai, India.

First edition: November 2011

10 9 8 7 6 5 4 3 2 1

Printed in the United States of America.

Contents

Figures and Tables

Figures

Tables

Foreword

Michael Montoya and
Rayna Rapp

As North American researchers invited to participate in a conference on the many biosocial issues raised in Latin American admixture studies, our curiosity and concern were both amplified and assuaged by the rich presentations and conversations that unfurled at "Racial Identities, Genetic Ancestry, and Health in South America" in 2009. Sponsored by the Wenner Gren Foundation and the British Academy, the excellent organizational work of Sahra Gibbon, Ricardo Ventura Santos, and Monica Sans brought us together outside the Americas, in London, and has kept us connected through their fine and multilingual editorial process that has produced the chapters you are about to read.

There is no tidy way to reflect upon the complexity and diversity of issues raised in this important volume. These chapters themselves reflect the fructuous, and often tense deliberations that were occasioned by the conference, one of whose aims was to decenter the European and North American hegemony on the biocultural, bioethical, sociocultural, historical, and political implications of genomics. However this meeting, like so many on our side of the Atlantic, also aimed to forward a conversation across the epistemological divides of the social and biological sciences. This book, therefore, reflects a consequential and ambitious agenda.

When North American researchers really engage the way that social and biological divides are being reconfigured by our South American colleagues, it becomes increasingly clear that many of our received, hegemonic ideas about "race," "genes," and "health" are wholly inadequate in large measure because they are refracted through the race debates in the United States. Still, the chapters by Monica Sans, Francisco Raúl Carnese, and Bernardo Bertoni, for instance, were at once troubling and reassuring. They were troubling because each in its own way wholeheartedly affirmed the science of human genetic variation, an affirmation that in the United States all too often marks a screen for unexamined or overt elision of the historical legacies of racial conflict and violence, masking how "admixture" was called into being and continues to evolve. It is apparent, for example, that the biparental DYS 199 locus and mitochondrial DNA techniques for characterizing populations deployed by some authors in this collection advance important technical debates about population structures in these specific contexts while at the same time addressing popular and often politically charged ideas about race/color and identity. We would have preferred that the problematics of haplogrouping, the ecological fallacy, or the nonconcordance of phenotype with genotype, nicely highlighted by Telma de Souza Birchal and Sérgio D. J. Pena, had been included in every chapter. At the same time, we learned a great deal about how the use of highly specific genomic techniques can be deployed with far greater nuance than what we too often note in the United States. Jose C. Flores, for example, points to the regional variations in admixture and diabetes susceptibility across Latin American countries, showing that when socioeconomic status is carefully added to the analytic variables, the impact of genomic ancestry-marker differences is dramatically reduced. And Bertoni insists on the complexity of epigenetic and regional environmental factors that make admixture studies partial, at best. Latin American analysts carry a power sensitivity, a context dependency, and an explicit recognition of the sociocultural, political, ethical, and historical responsibilities of the genetic sciences, which has much to teach US-based researchers.

Some of the chapters may seem at first to reproduce problematic racialized human difference. Yet they also, in sometimes strikingly opposite ways, deploy their genomic technologies in deliberate and visible accordance with the social exigencies of the peoples from whose bodies their genomic differences were derived. For example, some suggest, as Guilherme Suarez-Kurtz does, that specific pharmacologies might be made more population sensitive to underserved groups with the least expensive but diversity-sensitive biomarker tests; this is not a matter of genome testing but of population awareness at the most sophisticated biosocial level. Most importantly, the biogenetically trained authors' use of populations almost always appears in carefully qualified ways such that one rarely forgets that the population labels used by scientists are historically mediated descriptions, and that the matrices of biogenetic population difference were not intended to discreetly define groups: when biomarker testing is dense and large numbers of subjects are recruited, clinal distributions always appear. This is, once again, an important perspective, which North American scientists might ponder to good effect.

The editors have adroitly assembled contributions that not only individually but also in conversation point to the unending social consequences of slippery race labels in which social values attached to phenotype do not correspond to complex admixture measures at the genotypical level, for example, in essay by Peter Fry and in Birchal and Pena. Population differences were qualified in the body of each chapter or were qualified through the careful pairings of chapters by the editors. These important qualifications include: (a) a reference to a definition of race or ethnicity in its specific national and regional context—as Suarez-Kurtz tells us, the specific mix of African-derived and Euro-derived markers that underlie cultural accounts of "color"/ race in northern Brazil do not correspond to the biological findings in that region and certainly are not valid for describing "color" or racial admixture in the south; (b) a qualification that biogenetic differences are estimates only, sometimes based on large studies that inspire both statistical and social confidence, but at other times based on small representational studies whose sampling techniques are fragile, at best; and (c) that the genetic distinctions that can be estimated must be understood as variation along a continuous gradual geographically patterned clinal distribution of genetic variation. Such qualifications partially avoid the implicit typological and subspecies racializations common in gene-based studies that rely upon human variation where race and/or ethnicity are never qualified (Montoya 2011, pp. 56–57), and thus racialize as they make claims about disease or behavior or population structure. This collection affords a corrective to such unsituated genomic accounts; as Santos and Marcos Chor Maio write in this volume, "The narratives about the biohistory of the formation of Brazilian peoples that have been produced by the genome research along the lines of the 'Molecular Portrait of Brazil' are in tune with a deep seated sociological imagination that sees miscegenation as a positive, defining element in the identity of Brazil as a nation."

Thus these authors show us that "race" in Latin America can be prevented from operating as a purely biological form and instead rightly is configured in these chapters as an ontologically open cultural form with overlapping social, biopolitical, and molecular registers. Having engaged in the US-based debates surrounding genomics and race, we emphasize that this kind of analytical care is an important commitment among these Latin American researchers and should be highlighted for the work it does in setting standards for interdisciplinary collaborations on population difference and health. Researchers from the Northern Hemisphere have substantial methodological and theoretical lessons to learn from our colleagues who work in the south.

It could be said that, like some North American scientists who also claim a science that is responsible to their Latino or African ancestral communities, the South American scientists are positioning their claims as Latourian passage points for all others who desire to colonize the continent in well-known extractive material and semiotic maneuvers. However, such a label of social complicity would constitute an underappreciation of the technoscientific cultural forms being co-configured by the South American research enterprises on which these chapters report; we see genomic sciences applied toward local and regional questions of the apportionment of human diversity (Lewontin 1972; Bastos-Rodriguez et al. 2006; Watkins et al. 2003; Birchal and Pena this volume). This alone is a contextualizing move, but these genetic studies are further applied toward complicated issues of social inequity, prejudice, and intergroup misunderstanding. Thus, there exists in these papers virtually no intent to use an uninterrogated notion of "race" as a proxy for either the complex histories of colonization, slavery, and European and Asian migration to the continent; nor as a too-simple marker for what in the Unites States gets racially labeled as "health disparities." From debates about Brazil's "racial democracy" in the face of mobilized social movements for "affirmative action"; to the political push back from the right and left against these situating accounts and claims, the scientists whose work you are about to read articulate the political stakes that are widely held in the study of highly admixed populations. Latin America is a continent long known to be in a state of *mestizaje*, now molecularized under conditions of lively debate.

In sum, we remain both skeptical and optimistic of the transdisciplinary conjoining of the social and biological sciences in South America as they pertain to the deliberate and simultaneous use of science for sociopolitical and bioscientific ends. Like many conversations where "race," "identity," "genetics," and "health" are invoked, some of the tensions that surfaced in our meeting were the result of the classic conundrum of social definition—which admits history, politics, and power versus attribution of molecular difference—where biological divergences are marked as absolute (Montoya 2007). When the editors write in the introduction that "the region is characterized by proportions of genetic admixture resulting in wide variations in degree and types of disease susceptibility within and between national contexts," our North American–inflected pessimism worries that the laboratory scientists may be incautious by reductionistically racializing peoples and groups they study. Yet, the conference and this volume have also engendered some optimism not only because the geneticists often carefully qualified their taxonomies and assumptions in their papers, but also because there was a genuine dialogue between those trained in the biological sciences and those trained in the social sciences. In fact, many of us were mistaken for one another. This is a very good sign indeed.

References

Bastos-Rodrigues, L. et al. (2006) The genetic structure of human populations studied through short insertion-deletion polymorphisms. *Annual Review of Human Genetics,* 70, pp. 658–665.

Lewontin, R. C. (1972) The apportionment of human diversity. *Evolutionary Biology,* 6, pp. 381–398.

Montoya, M. (2007) Bioethnic conscription: Genes, race and mexicana/o ethnicity in diabetes research. *Current Anthropology,* 22(1) pp. 94–128.

Montoya, M. (2011) *Making the Mexican diabetic: Race, science and the genetics of inequality.* Berkeley: University of California Press.

Watkins, W. S. et al. (2003) Genetic variation among world populations: Inferences from 100 *Alu* insertion polymorphisms. *Genome Research,* 13, pp. 1607–1618.

Acknowledgments

The publication of this interdisciplinary collection is the fruit of different collaborations and the efforts of many persons who have both individually and collectively helped realize this timely and important book. The conversations that were the seed of this collection began in February 2009 as part of a unique interdisciplinary event at University College London (UCL) that brought together 25 participants from South America, the United States, and the United Kingdom to address and consider the social and cultural implications of new developments in the field of population genetics for health and identity. This event was generously funded by a workshop grant from the Wenner Gren Foundation and the British Academy UK-Latin America/Caribbean Link Programme. We also benefited from the support and help of colleagues and staff in the Department of Social Anthropology at UCL who agreed to host the public symposium and the workshop event. We would in particular like to thank Nanneke Redclift, then head of the Anthropology Department at UCL, whose support and enthusiasm for the event was unstinting and whose characteristically generous, intelligent, and engaged commentaries helped nurture the crossdisciplinary dialogues and exchanges that took place. We are also extremely grateful to the other participants who took part in the workshop in different ways whose collective participation helped enrich and contribute to the discussions but whose presentations are not all included in this collection. In particular, we would like to thank Peter Wade, Alicia Caratini, Flávio Gordon, Steve Humphries, Richard Ashcroft, David Skinner, and Mark Thomas. We are also grateful to the seamless behind-the-scenes activities of Anne Rudolph who was on call prior, during, and after to facilitate the smooth flow of the workshop.

We would particularly like to thank the Biblioteca Nacional in Rio de Janeiro who granted permission to reprint the map of South America, which forms part of the front cover. We are also grateful to Gary Hutson whose visual acumen helped guide and inform the choice of images for the design.

SAHRA GIBBON
RICARDO VENTURA SANTOS
MÓNICA SANS

Introduction

*Sahra Gibbon, Ricardo Ventura Santos, and
Mónica Sans*

In 2007, a daily newspaper in Salvador in the state of Bahia, Brazil, published a wonderfully creative image by the Brazilian artist and cartoonist Cau Gomez.[1] Simultaneously, it evokes and references the well-known drawing by the Renaissance artist Leonardo de Vinci's Vitruvian man and the debates about race and genomics that have become frequent in Brazil and other South American countries in the first decade of the twenty-first century. Like other countries of the world, new technological developments in the field of genomics have provoked intense debates about the historical formation of the Brazilian society and the construction of its national identity. As the chapters in this book demonstrate, such debates have crossed the boundaries of the laboratory in various national and transnational contexts in South America and are intersecting with and being informed by public discourse and political practice with diverse consequences.

If the man designed by da Vinci is a canon of classic beauty, with what was regarded by the artist as perfect proportions between the diverse parts of the body, in the Brazilian version it is not only the geometry of the human body highlighted. With the title "Genetic Composition (the DNA of the Brazilian People)," what is being made explicit is not the perfect proportions that da Vinci's drawing emphasized, but "imperfections." The body parts of this Vitruvian Brazilian man are disproportionate, as if they result from poor health conditions, such as undernutrition, tuberculosis, worms, and rickets: the head is disproportionately large, the bones of the ribcage are visible under the thin skin, the stomach is swollen, and the limbs are thin and fragile. Moreover, the message here is that genes do not explain everything about this specific human complexity: the percentages of biological ancestry only combine to make 100 percent when famine, unemployment, and illiteracy are added. More visibly

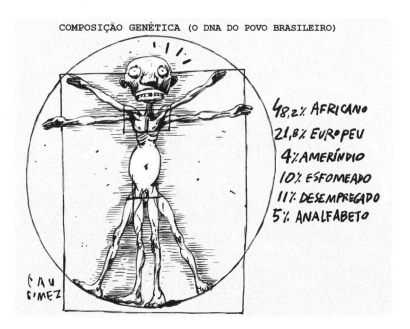

COMPOSIÇÃO GENÉTICA (O DNA DO POVO BRASILEIRO)

48,2% AFRICANO
21,8% EUROPEU
4% AMERÍNDIO
10% ESFOMEADO
11% DESEMPREGADO
5% ANALFABETO

CAU GOMEZ

Figure 0.1   Cartoon "Composição genética (O DNA do povo brasileiro)" (Genetic composition—the DNA of the Brazilian population—48.2 percent African, 21.8 percent European, 4 percent Amerindian, 10 percent starving, 11 percent unemployed, 5 percent illiterate).

Source: Cau Gomez (reproduced with permission).

African than European, with a few pinches of Amerindian genes, the Vitruvian Brazilian man is a mestizo/*mestiço*, with his eminently biocultural genetic composition that includes as much his biological ancestry as his "socioeconomic and political DNA."

It would be difficult to imagine a more appropriate image than that drawing by Cau Gomez to introduce this book. Here an image that references the Renaissance quest to understand the relationship between humans and nature is reworked to raise profound and deeply important questions about the relationship between racial identities, genetics, society, and politics that speaks not only of Brazil and South America, but also the meaning and utility of genetics in the contemporary world as a whole.

In the latter half of the twentieth century the ability to examine human genetic variation with greater specificity and detail has challenged entrenched ideas about strong or definitive biological differences between humans (Lewontin 1972; Cavalli-Sforza 1994). At the same time, questions remain about how novel genetic knowledge and techniques for examining population variation are informing questions of "difference" and "sameness" and the social or ethical challenges of medical interventions that will arise from this knowledge (Duster 2003, 2005; Ashcroft 2006; Ellison and Goodman 2006; Pálsson 2007; Fujimura et al. 2008; Marks 2008). These developments are evolving very rapidly in a diverse set of social, cultural, political, and institutional contexts, informing, mobilizing, and impacting upon preexisting individual and collective histories of ancestry, colonization, and identity (Brodwin 2005; Salzano 2004; Montoya 2006; Palmie 2007; Pálsson 2007; Fulwilley 2007, 2010; Santos et al. 2009; Wade 2010). There are key questions here for those concerned with the global context of developments in genomic knowledge and technology.

One of these is the pressing need to develop appropriate theoretical and conceptual tools, which can speak to and across disciplinary divides in the sciences and social sciences, helping to ensure that new genetic knowledge does not get caught up in "old" racial terms (Ellison and Goodman 2006; Skinner 2006; Goodman 2007; Pálsson 2007; Weiss 2007). At the same time, these developments provide an opportunity for new forms of cross-disciplinary research, which can more usefully address and examine the social and scientific issues at stake. However, the analytic capacity to evaluate the implications of this process has not only been limited by disciplinary divisions or the need for theoretical and methodological innovation, but also by a somewhat narrow focus on Europe and more specifically North America. While there has been great deal of genetic research in South America on "admixture" (see Bortolini and Salzano 1996; Sans 2000), currently much less is known about the social and cultural context and consequences of novel genetic knowledge relating to human biological variation in this and other areas of the world, including South America.

This book arises from the meeting that took place at University College London in 2009 in an event generously funded both by the Wenner Gren Foundation and the British Academy. The event brought together participants from Argentina, Brazil, Colombia, Uruguay, the United States, and the United Kingdom with the explicit aim of examining the scientific and cultural meaning of new genetic understanding of human biological difference in South America from different disciplinary perspectives. The experience and research interests of participants rested on examining and reflecting on the implications of novel developments in the field of genetics for understanding history, identity, and health in the South American region. Disciplinary expertise was extremely varied, including anthropologists, sociologists, historians, geneticists, and bioethicists. As a result, there was more than a degree of uncertainty (at least on the organizers' parts) about how the conversations would evolve over the three-day event, which included a public symposium and an invitee-only workshop. While there were heated discussions on both sides, what emerged was a period of intense and productive engagement that in turn generated novel and unexpected understandings, new questions and, for the majority, ample motivation for continuing to develop and amplify both cross-disciplinary and interdisciplinary inquiry. This book represents an opportunity to bring these debates and the issues that confront those working in South America to a wider audience.

Race, Racial Identities, and Genomics: Old and New Debates

The unexpected revival of discussion and interest around race in the context of genomic research has been highlighted in two recent edited collections that mostly focus on the North American and European contexts. In the collection edited by B. Koenig et al., entitled *Revisiting Race in a Genomic Age,* it is noted that "contrary to…expectations and hopes, post-genomic science has revived the idea of racial categories as proxies for biological differences" (2008:1). While in their collection *What's the Use of Race? Modern Governance and the Biology of Difference,* I. Whitmarsh and D. S. Jones note that "what is unexpected, however, is the way in which the relevance of race as a social, legal and medical category has been reinvigorated by science, especially genetics" (2010:2).

These new perspectives on race in the context of emerging genetic technologies have concentrated on a diverse set of themes. The way that race is and should be used in evolving spheres of governance as this concerns emerging genetic technologies are the focus for the collection by Whitmarsh and Jones (2010). Highlighting both the "promise" and "dangers" at stake, they point to the ways that race and genetics are being incorporated into three spheres of governance as this relates to "ruling," "knowing," and "caring," showing how both old and new categories, political practices, and institutions are given novel emphasis and meaning (2010). Koenig et al. (2008) identify four issues that reflect the specificity and novelty of these contemporary configurations. A broader discussion of these four themes outlined below provide a starting point for thinking about how genetics is impacting on race or racial identities and the potential for what has been described as the "molecular re-inscription of race" (Duster 2006; Bolnick 2008).

The first theme relates to the completion of the sequencing of the human genome in the early twenty-first century, which has been widely recognized as commencing the "genomic" age and instituting a shift from relatively limited gene-hunting research to whole-genome analysis. Here despite much old and new debate about the extent to which patterns of human genetic variation (variously defined) map to race (see for instance Risch et al. 2002; Bamshad and Olsen 2003; REGWG 2005), race and ancestry have nevertheless become central "research tools" in genomic science. They are an essential part of the technological apparatus of an evolving and expanding field of population genetic research (Bolnick et al. 2007; Koenig et al. 2008).

The second issue is the way that academic race and genetic research is now entering the market place. This has been nowhere more apparent than in the commercial investment in pharmacogenetics (Ashcroft, 2006) and the rapid expansion and growth in the United States of direct to consumer commercial genetic testing for disease risk or for ancestry testing (Bolnick et al. 2007). This commercialization has also become apparent in the high-profile marketing of the first FDA approved "race-based drug" BiDiL that was targeted at African Americans in the United States; a development that has become the focus for heated debate about health rights and the scope for genetic research to biologically reinscribe social categories such as race (Kahn 2005).

The developments and debates around the marketing of BIDIL also make it very easy to see, in the US context at least, how research on race and genetics has now become directly caught up with the politics of identity, which is the third thematic issue identified by Koenig et al. (2008). The issue of genetic knowledge or technology and identity has been of long-standing concern in the social sciences (see Gibbon and Novas 2008). Yet these developments in the science of human biological variation are playing out in diverse contexts and with, as yet, still to be assessed consequences as a discourse of rights, and the politics of "underserved" populations become central to investment and interest in race-inflected genetic research framed as "social justice" (Lee 2008; Epstein 2007). In the United States, they are also being informed by the expansion of commercial ancestry testing that has become an essential part of the genealogical "roots" industry among African American and Native American communities (Nelson 2008).

Finally, Koenig et al. point to the way today genetic research on race increasingly takes place in a medical context that moves beyond as well as recalibrates interest in the history of human migration to more medically focused population genetic research. Yet for Koenig et al., it is the twin and somewhat contradictory goals of "personal medicine" and the need to address "health disparities" in the United States that is they suggest propelling "social investment in genetic research on human population variation" (see also Shields et al. 2005).

Recent collections, such as those discussed above by Koenig et al. (2008) and Whitmarsh and Jones (2010), bring to the fore an important set of analytical perspectives in addressing the evolving interface between race, racial identities, and contemporary genetic science. Both books highlight the way that a range of different and still-emerging genetic "technologies" (pharmacogenomics, admixture mapping, genome-wide association studies [GWAS], Single Nucleotide Polymorphism [SNPs]) are central driving forces in this context. Yet the interest in both cases is very clearly the historical context, geographic, cultural, and political particularities of North America and Europe over the last few decades. Our goal here is to extend these fledgling comparative studies that include different disciplinary studies from within South America, to create a broader interdisciplinary perspective from diverse social, political, and cultural contexts to shed new light on how these debates are evolving in arenas outside the United States and Europe.

A View from the South: Race Identities and Genomics

In the introduction of her well-known book on eugenics in Latin America, historian of science Nancy Stepan discusses why, in the mid 1980s, there were so little historical research focusing on the topic in that particular region. She argued that, up to that moment, there was a lack of interest on the history of eugenics in Latin America because it was perceived "as a consumer and not as a contributor of ideas, and a fairly passive one at that" (Stepan 1991:3).

Stepan's work, with its emphasis on Latin America, has helped to show that, although eugenics was a commonplace in Western thought, it could not be understood as a homogenous, unified scientific branch of knowledge, defined by shared interests and goals on a global scale (see Adams 1990). Rather, as pointed out by Bauman, it was the "legitimate offspring of the modern spirit, of that urge to assist and speed up the progress of mankind toward perfection that was throughout the most prominent hallmark of the modern age" (1991:33). Over the past two decades, the pioneering constructivist perspective put forward by Stepan has stimulated a number of investigations on the history of eugenics in Latin America, which have further indicated that eugenics was a highly social activity entrenched with the values of the societies in which it was produced and practiced (see Hochman et al. 2010).

The intellectual trajectory of history of science, as it concerns the investigations on eugenics in recent decades, provides a good starting point for the construction of a critical contemporary agenda on issues related to racial identities and genetics. If Stepan's work highlighted the need to attend to the particularities of Latin American histories and contexts, current developments in the field of human genetic variation are demanding an equal attention to the variety and specificity of the region in thinking about the consequences of these developments for identity, health, and nationhood. Such a "view from the south" might help us to better understand the contingencies and sociocultural and political specificities involved in the production and application of new biological knowledge and technologies in areas of the world away from the North American—European axis.

Thus our main argument in this volume is that South America provides a unique and important arena for examining the social and ethical issues raised by these developments. As a whole, the region is characterized by different proportions of genetic admixture (Bortolini and Salzano 1996; Sans 2000; Salzano and Bortolini 2002), resulting in wide variations in degree and type of disease susceptibility between and within national contexts (Chakraborty et al., 1999). In this sense, South America not only provides an important means to include and account for genetic heterogeneity, but also it is argued that genetic research in the region is vital to the mapping and identification of disease-susceptibility gene variants, the development of pharmacogenetic targets, and understanding the interaction between biological and environmental contribution to disease occurrence and drug response (Chakraborty and Weiss 1986; Gonzalez-Burchard et al. 2005; Suarez-Kurtz 2005; Pena et al. 2011).

Against this long history of research and contemporary interest in the genetic composition of Latin American populations, the sociopolitical meaning of race and ethnicity in this area is diverse, both historically unstable and contested. As national identities in Latin American countries have been constructed in racial terms we can see that with "pressures to transform the role of the nation-state has come a resurgence of ethnic and racial movements and across national borders" (Applebaum et al. 2003:21). Thus, the ways that social and political histories of ideas associated with race, nation, citizenship, and ethnicity have evolved in this setting (Stepan 1991; Wade 1993, 2002, 2010; Applebaum et al. 2003), provide a uniquely important contrasting historical and sociopolitical arena, outside Europe and the North America, for understanding the diverse public/science interfaces at issue in the evolving field of genomics.

Racial Identities, Genetic Ancestry, and Health in
South America: Thematic Overview

The book is divided into three sections offering perspectives on the challenges and opportunities of interdisciplinary research and practice, new developments in genomics as this relates to different fields of health in South America, and in the final section, the relationship between genetic knowledge, history, and identity within different national context. Our case studies refer to a limited set of South American countries (Argentina, Brazil, Colombia, and Uruguay), but we believe that the issues raised are, in general, representative of what is at stake in the region as a whole. Although in each section the notion of "racial identity" is more or less explicitly addressed, it is nonetheless constantly revealed in the South American region as multilayered and complex sometimes hidden, disguised, and unspoken and at other times brought into explicit discourse and practice (see also Applebaum et al. 2003; Wade 2010).

Considering the scope of interdisciplinary research, the chapters in the opening section of the book illuminate the specificity of the South American context. The chapters by Ricardo Ventura Santos and Marcos Chor Maio and also Telma S. Birchal and Sérgio D. J. Pena focus in different ways on how the intersections between genetic research and what is described as "the paradox of racial identity" in the region have intersected in the doing of and in the reception to genetic research. While not directly discussing "biocultural anthropology," there is an implicit emphasis in the first two chapters about how interdisciplinary perspectives are central to emerging fields of genetic research on human difference linked to the "political" consequences of such research. While very clearly shedding light on the particularities of the South American context, where actual genetic mixing and cultural ideas of being mixed race continue to powerfully inform these developments, the articles also beg the question if the Brazilian cultural context is unique or part of a wider set of regional cultural and social practices. At the same time, the implications of an as yet uncommon but productive partnership between a geneticist and a philosopher, in chapter 3 by Pena and Birchal, suggest that there are implications that go far beyond the Brazilian national scenario. Their discussion of the ethical issues related to the use of knowledge on the biological/ genetic constitution of populations in the definition and implementation of public policies reveals both the productivity and ongoing need for reflective interdisciplinary discussions across diverse national and transnational arenas.

Andres Barragan expands and complexifies these issues by examining the "paradox of racial identity" in the Colombian context. Placing the legacy of biopolitical histories of transnational genetic research in Latin America centre stage in understanding contemporary response to expanding fields of genetic science in Colombia, his chapter illuminates the "sociopolitical frictions" that arise in Colombia at the "meeting point" between expanding scientific interest in mestizo or "admixed" populations, shifting ethnic politics and local-global ecological agendas, issues also addressed in the chapter by Santos and Maio for Brazil. Here too, as in the previous chapters, we can see a high degree of polarization between communities of activism and different disciplinary perspectives; a situation that, as the authors in this section suggest, demands novel strategies of engagement and communication between and within different academic and activist participants collectives.

The second section of the book brings the health field more clearly into view. It reflects on how seemingly novel and more long-standing areas of health-care interventions in South America are being refracted through and in response to the development of genetic science, as well as debates and concerns about doing so. The notion of "genetic admixture" takes centre stage here as this relates to the application of genetic knowledge to health care in the South American region. While Bernardo Bertoni provides this section with a starting point for discussing recent technical developments in understanding and utilizing genetic admixture studies, other chapters point to the challenges and opportunities of doing so in South America. Guilherme Suarez-Kutz's chapter powerfully illustrates the danger of imposing targeted pharmacogenomic interventions for different disease conditions in Brazil developed on the basis of foreign racial categories and biomedical protocols. His chapter raises profound questions about the scope of such science and points to the undesirable consequences for health policies if not properly attuned to the specificities of "genetic admixture" not only in South America but in other seemingly "less" obviously admixed arenas also. The chapter by J. C. Florez examining the relationship between non-European ancestry and type 2 diabetes based on research in Mexico and Colombia raises questions about the need to incorporate social and economic status in genetic admixture studies aimed at understanding the relationship between ancestral genetic markers and the high incidence of diabetes in different regions of South America. Echoing the chapters in the first part of the book, as well as the powerful critique visualized in Cau Gomez's pictorial representation, the article makes a powerful case for further cross-disciplinary analysis in ways that can incorporate factors such as social and economic inequities into genetic analysis.

Other studies suggest that there is still much work to be done in the work of integrating such factors, not simply as variables in research findings but as lived social, cultural, and historical realities that inter-act and themselves directly inform the biosocial components of health differentials in diseases such as diabetes and other common conditions (Montoya 2006; Krieger 2010). Peter Fry's chapter concludes this section reflecting on the way public health awareness and response to a sickle cell has developed in Brazil. As one of the first diseases to be identified as genetic, sickle cell marks an important development in the history of molecular medicine. As other work has shown in the United States, it is a condition where the relationship between disease and race has worked to mutually inform one another such that, in the United States, sickle cell disease became in the 1970's "the defining characteristic of being black" (Tapper 1998). Peter Fry's chapter examines the emergence of awareness of the disease with black activism in genetically "admixed" Brazil, show-ing how this uneasy alliance operates in tension with, on the one hand, political discourse that is committed to promoting Brazil as a "racial democracy" and, on the other hand, emerging scientific discourse that constitutes the individual as a genetic "mosaic."

The third and final part of the book reflects on the different ways that genetic knowledge has in the past and continues to be caught up with nation building and the reproduction of different "national" identities across comparative regions of South America. The chapters in this section draw attention to the ways in which a variety of different tools and tech-nologies of genetic inquiry including census categories, forensics, records of family marriage, tissues samples, or SNPS have been mobilized in pur-suit of understanding population history or in the case of one article in this section, recover identities that were violently and brutally suppressed. In the first chapter, Francisco R. Carnese et al. examine the heterogeneity of the population of the Argentinean capital city, Buenos Aires. Although showing that in the same city, differences of origin can be found, their conclusion, based on genetic markers related to place of birth and records of family marriage, is, that the metropolitan region cannot be generically described as a "melting pot." The second chapter in this section by Mónica Sans examines the changing concept of national identity in Uruguay over the last few decades and how this is related to individuals self-ascription to different "ethnic/racial" groups. Comparing and analyzing historical sources, census, and genetic data, a particular population panorama is revealed. This is shown to be not necessarily or always congruent with popular Uruguayan beliefs of national identity that, it is argued, have tended to rest on an idea of a homogenous population with European origins that includes extermination of indigenous "native" populations. In both these chapters what is being highlighted in different yet compli-mentary ways is the "gap" between genetic understanding and popular perceptions of identity variously configured through the lens of nation or perceptions of origins. From a different yet complementary perspective, Ann Morning's comparative perspective on the way that census categories were applied globally in the 2000 census provides an intriguing glimpse of the extent to which ambiguity and diversity characterizes how race, ethnicity, and concepts of national citizenship get deployed in census practices (Morning 2008). As Catherine Bliss points out, we need to examine how US census categories travel in transglobal markets of genomic research and "in places that hold citizenary to be racialized in different ways" (2010:12). The chapters by Sans and Carnese et al. may be seen as part of an ongoing debate related to the collection of information on race/ethnicity in the context of censuses and how this relates to the practice and politics of genetics research itself.

The final chapter in the book while also contributing to the discussion of how genetics is linked to national identity also, sheds light on the issue of ethics and bioethics in genetic research particularly as this relates to different aspects of national history and identity. The account in Victor Penchaszadeh's chapter of the use of forensic genetics in the "recovery" of the identity of children of the disappeared by the military regime in Argentina in the 1970s and the 1980s is both moving and disturbing. Yet it also points to the scope of not only forensic genetics but the need for interdisciplinary research also. These approaches may be particularly important in the pursuit of uncovering national histories and identities in other places in South and Central America where the military state practice of "disappearing" so called dissidents and illegally adopting children has not been uncommon. The accounts in this chapter also raise important questions about the way that contemporary cultural meanings of genetics as being the "essence" of identity and seemingly older "Lamarckian" notions of heredity intersect in the ways that people whose identities have been "recovered" in this way make sense of new genetic knowledge. What is clear in all the final three chapters in this section is that the use of genetic science is central to the constitution of national identities and histories, opening up a new ethical space that reflects and informs the meaning of genetic science in the South American context.

Concluding Remarks

The chapters in this book provide a window on to the historical, social, and political context in which new genetic research on human genetic variation is taking place in South America. While it is undoubtedly the case that such research is now global in ways that generate novel trans-national modalities for genetic science (Ong and Collier 2005), it is also clear that these intersect with historically defined cultural and political parameters that have and continue to constitute race and racial identities as well as genomic knowledge in particular ways.

The various chapters, in coalescing social and scientific studies of a range of different conditions across national contexts, bring to the fore-front some of the tensions at stake in considering how health, genetic knowledge, racial identities, ethnic politics, and human rights are being configured in South America. While some tensions would seem generic to emerging transnational fields of genetic science and medicine (between novelty and/or the reenforcement of older cultural categories for instance), what is striking is the way developments in South America related to genetic research and health are informed by the "truth of admixture." It is a context that both displaces and reframes debates that have emerged in the US context regarding the "molecularization of race" and genetic research concerned with understanding and addressing health dispari-ties. The articles outlined here show how such debates do not easily or directly translate into the South American arena. Here particular histo-ries of colonization, ethnic/racial or social classification, as well as the very different emergence of indigenous and black rights–based move-ments that crosscut ongoing health and social inequities, inform and transform these debates revealing a particularity in the way health, racial and racialized identities, and genetic ancestry are being configured.

There is much research still to be done in this regard not least in examining the biopolitics of evolving transnational genetic research trajectories as they connect and increasingly potentially shift the power dynamics in and between the global south and north. There are also questions about how the widening paradigms of genomic research, which now increasingly incorporate "epigenetic" frames of reference that include the role of intracellular and extracellular environments, will take account of the sociocultural realities that Cau Gomez so astutely alludes to. The extent to which poverty, inequities, and racism become not only a justification for genetic research but also an integral part of understand-ing disease etiology alongside genomics, remains an elusive yet sought after objective for those who consider this to be an interdisciplinary imperative (Ellison and Goodman 2006; Krieger 2010). With a strong emphasis on *mestizage/ mestiçagem* in the dynamic processes of con-stituting national identities, an expanding scientific and technological basis for the development of genomic research, and also a political will across many national contexts to challenge entrenched problems, such as poverty and inequities in health, it is tempting to suggest that South

America is well placed to contribute to the challenge of truly interdisciplinary inquiry. Given the renewed interest and polarization of positions on race and racial identities in many international arenas the stakes could not be higher.

Note

1. See Cau Gomez's webpage for further information at www.caugomez.br.

References

Adams, M. B. (1990) Toward a comparative history of eugenics. In: Adams, M. (ed.) *The Wellborn Science: Eugenics in Germany, France, Brazil and Russia.* New York: Oxford University Press, pp. 217–231.

Applebaum, N. et al. (2003) (eds.) *Race and Nation in Modern Latin America.* Chapel Hill: University of Carolina Press.

Aschroft, R. (2006) Race in medicine: From probability to categorical practice. In: Ellison, G. and Goodman, A. H. (eds.) *The Nature of Difference: Science, Society and Human Biology.* Boca Raton, Florida: CRC Press, Taylor & Francis Group, pp. 135–157.

Bamshad, M. J. and Olsen, S. E. (2003) Does race exist? *Scientific American,* 289, pp. 78–85.

Bauman, Z. (1999) *Modernidade e ambivalência.* Rio de Janeiro: Jorge Zahar.

Bliss, C. (2010) Census, race and genomics. *Anthropology News,* May, pp. 9–10.

Bolnick, D. A. (2008) Individual ancestry inference and the reification of race as a biological phenomenon. In: Koenig, B. A. et al. (eds.) *Revisiting Race in a Genomic Age.* Piscataway, NJ: Rutgers University Press, pp. 70–85.

Bolnick, D. A. et al. (2007) The science and business of genetic ancestry testing. *Science,* 318 (5849), pp. 399–400.

Bortolini, M. C. and Salzano, F. M. (1996) MtDNA diversity analysis in Amerindians and other human populations—how different are they? *Brazilian Journal of Genetics,* 19, pp. 527–534.

Brodwin, P. (2005) "Bio-ethics in action" and human population genetic research. *Culture, Medicine and Psychiatry,* 29, pp. 145–178.

Cavalli-Sforza, L. I. et al. (1994) *The History and Geography of Human Genes.* Princeton: Princeton University Press.

Chakraborty, B. M. et al. (1999) Is being Hispanic a risk factor for non-insulin-dependent diabetes mellitus? *Ethnicity & Disease,* 9, pp. 278–283.

Chakraborty, R. and Weiss K. M. (1986) Frequencies of complex diseases in hybrid populations. *American Journal of Physical Anthropology,* 70, pp. 489–503.

Duster, T. (2003) Buried alive: The concept of race in science. In: Goodman, A. H. et al. (eds.) *Genetic Nature/Culture: Anthropology and Science beyond the Two-Culture Divide.* Berkeley: University of California Press, pp. 258–277.

Duster, T. (2005) Race and reification in science. *Science*, 307(18), pp. 1050–1051.

Ellison, G. T. H. and Goodman, A. H. (eds) (2006) *The Nature of Difference: Science, Society and Human Biology.* Boca Raton, FL: Taylor and Francis.

Epstein, S. (2007). *Inclusion: The Politics of Difference in Medical Research.* Chicago: University of Chicago Press.

Fujimura, J. H. et al. (2008) Introduction. Race, genetics and disease: Questions of evidence, matters of consequence. *Social Studies of Science*, 38(5), pp. 643–656.

Fullwilley, D. (2007) The molecularization of race: Institutionalizing human difference in pharmacogenetics practice. *Science as Culture*, 16(1), pp. 1–30.

———. (2010) Revaluating genetic causation: Biology, economy and kinship in Dakar, Senegal. *American Ethnologist*, 37(4), pp. 638–661.

Gibbon, S. and Novas, C. (2008) *Biosocialities, Genetics and the Social Sciences. Making Biologies and Identities.* London: Routledge.

Gonzalez-Birchard, E. et al. (2005) Latino populations: A unique opportunity for the study of race, genetics, and social environment in epidemiological research. *American Journal of Public Health*, 95(12), pp. 2161–2168.

Goodman, A. (2007) Towards genetics in an era of anthropology. *American Ethnologist*, 34(2), pp. 227–230.

Hochman G. et al. (2010) Paths of eugenics in Brazil: Dilemmas of miscegenation. In: Alison B. and Philippa L. (eds.) *The Oxford Handbook of the History of Eugenics.* New York: Oxford University Press, pp. 493–510.

Kahn, J. (2005) Misreading race and genomics after BiDil. *Nature Genetics*, 37, pp. 655–656.

Koenig, B. et al. (eds.) (2008) *Revisiting Race in a Genomic Age.* Piscataway, NJ: Rutgers University Press.

Krieger, N. (2010) The science and epidemiology of racism and health: Racial/ethnic categories, biological expressions of racism, and the embodiment of inequality—an ecosocial perspective. In: Whitmarsh, I. and Jones, D. S. (eds.) *What's the Use of Race? Modern Governance and the Biology of Difference.* Cambridge, MA: MIT Press, pp. 225–259.

Lee, S. S. (2008) Racial realism and the discourse of responsibility for health disparities in a genomic age. In: Koenig, B. A. et al. (eds.) *Revisiting Race in a Genomic Age.* Piscataway, NJ: Rutgers University Press, pp. 342–358.

Lewontin, R. C. (1972) The apportionment of human diversity. *Evolutionary Biology*, 6, pp. 381–398.

Marks, J. (2008) Race: Past, present, and future. In: Koenig, B. A. et al. (eds.) *Revisiting Race in a Genomic Age.* Piscataway, NJ: Rutgers University Press, pp. 21–28.

Montoya, M. (2006) Bioethnic conscription: Genes, race and mexicana/o ethnicity in diabetes research. *Cultural Anthropology*, 22(1), pp. 94–128.

Morning, A. (2008) Ethnic classification in global perspective: A cross-national survey of the 2000 census round. *Population Research and Policy Review*, 27(2), pp. 239–272.

Nelson, A. (2008) Bio science: Genetic ancestry testing and the pursuit of African ancestry. *Social Studies of Science*, 38, pp. 759–783.

Ong, A. and Collier, S. J. (eds.) (2005) *Global Assemblages: Technology, Politics, and Ethics as Anthropological Problems.* Malden and Oxford: Blackwell Publishing.

Palmie, S. (2007) Genomics, divination and racecraft. *American Ethnologist,* 34(2), pp. 205–223.

Pálsson, G. (2007) *Anthropology and the New Genetics.* New York: Cambridge University Press.

Pena S. D. et al. (2011) The genomic ancestry of individuals from different geographical regions of Brazil is more uniform than expected. *PLoS One,* 6(2), e17063.

REGWG (Race, Ethnicity, and Genetics Working Group) (2005) The use of racial, ethnic, and ancestral categories in human genetics research. *American Journal of Human Genetics,* 77, pp. 519–532.

Risch, N., et al. (2002). Categorization of humans in biomedical research: Genes, race and disease. *Genome Biology,* 3(7) http://genomebiology.com/2002/3/7/comment/2007.

Salzano F. M. (2004) Interethnic variability and admixture in Latin America—Social implications. *Revista de Biologia Tropical,* 52, pp. 405–415.

Salzano, F. M. and Bortolini, M. C. (2002) *The Evolution and Genetics of Latin American Populations.* Cambridge: Cambridge University Press.

Sans, M. (2000) Admixture studies in Latin America: From the 20th to the 21st century. *Human Biology,* 72, pp. 155–177.

Santos, R. and Maio, M. C. (2006) Race, genomics, identities and politics in contemporary Brazil. *Critique of Anthropology,* 24(4), pp. 347–378.

Santos, R. V. et al. (2009) Color, race and genomic ancestry in Brazil: Dialogues between anthropology and genetics. *Current Anthropology,* 50(6), pp. 787–819.

Shields, A. et al. (2005) The use of race variables in genetic studies of complex traits and the goal of reducing health disparities. *American Psychologist,* 60, pp. 77–103 .

Skinner, D. (2006) Re-thinking "race" and science; teaching the biological and the social. In: Farrar, M. and Todd, M. (eds.) *Teaching "Race" in the Social Sciences: New Contexts, New Approaches.* University of Birmingham, Sociology, Anthropology and Politics (C-Sap) Birmingham: The Higher Education Academy Subject Network.

Stepan, N. L. (1991) *The Hour of Eugenics: Race, Gender and Nation in Latin America.* Ithaca and London: Cornell University Press.

Suarez-Kurtz, G. (2005) Pharmacogenomics in admixed populations. *Trends in Pharmacological Sciences,* 26(4), pp. 196–201.

Tapper, M. (1998) *In the Blood: Sickle Cell Anemia and the Politics of Race.* Philadelphia: University of Pennsylvania Press.

Wade, P. (1993) *Blackness and Race Mixture: The Dynamics of Racial Identity in Colombia.* Baltimore: John Hopkins University.

———. (2002) *Race, Nature, Culture: An Anthropological Approach.* London: Pluto Press.

Wade, P. (2010) *Race and Ethnicity in Latin America.* 2nd edition. London: Pluto Press.

Weiss, K. (2007) On babies and bathwater. *American Ethnologist,* 34 (2), pp. 242–244.

Whitmarsh, I and Jones, D. S. (eds.) (2010) *What's the Use of Race: Modern Governance and the Biology of Difference.* Cambridge, MA: MIT Press.

Part I

Interdisciplinary Research and the Paradox of Racial Identity

1

Anthropology, Race, and the Dilemmas of Identity in the Age of Genomics: A View from Brazil

Ricardo Ventura Santos and
Marcos Chor Maio

Introduction

It has been noted by many authors that the "new genetics" (or genomics) is having a huge impact on the most diverse areas of the contemporary world, creating a technical and cultural revolution involving genes that is changing technologies, institutions, practices, and ideologies (Lippman, 1991; Rabinow, 1992; Haraway, 1997; Goodman et al., 2003; Santos and Maio, 2004). The knowledge and technologies derived from the new genetics do not just lend new dimensions to the biological, cultural, and social loci in the near surrounds of individuals; they also reshape macrosocial, historical, and political relationships of a far broader scale. Anthropologist Paul Brodwin (2002) is categorical in his views on the relationships between the development of genetic technologies, society, and the construction of social identities in the contemporary world. As genetics earns greater prestige, historically recognized standards of identity may gain further legitimacy or be overruled by the results of DNA sequencing, or there might emerge new propositions that had not previously been socially acknowledged.

This is just such the state of affairs in the current-day relationship between race and genomics. In his book *Against Race*, Paul Gilroy states that anybody who wishes to consider the elements that have influenced what he calls the "crisis for raciology" should pay special attention to genomics. As he puts it, "[The] distance [of genomics] from the older versions of race-thinking that were produced in the eighteenth and nineteenth centuries underlines that the meaning of racial difference is itself being changed, as the relationship between human beings and nature is reconstructed by the impact of the DNA revolution and of the technological developments that have energized it" (Gilroy, 2000, pp. 14–15; also see 1998).

In advocating a "deliberate and self-conscious renunciation" of race as a method for categorizing and dividing humanity (p. 17), Gilroy stresses that the biotechnology revolution makes it necessary for us to alter our understanding of concepts like race, species, embodiment, and human specificity. In other words, it requires us to reconceptualize the relationship between ourselves, our species, our environment, and our notion of life: "We need to ask, for example, whether there should be any place in this new paradigm of life for the idea of specifically racial differences" (p. 20).

Underlining the "utopian tone" of his argument (p. 7), Gilroy recognizes that his radical "antirace" posture may compromise or hamper (or even betray) those groups whose legitimate and even democratic claims rest upon forms of identity that have been built up at great cost, based on categories imposed by their oppressors (p. 52). Race and its by-products constitute one such set of categories. As Gilroy sees it, to abandon "race" means to sever a long-standing historical chain by breaking one link: "On the one hand, the beneficiaries of racial hierarchy do not want to give up their privileges. On the other hand, people who have been subordinated by race-thinking and its distinctive social structures... have for centuries employed the concepts and categories of their rulers, owners, and persecutors to resist the destiny that 'race' has allocated to them" (2000, p. 12).

What we intend to discuss in this article is precisely this "articulation" between beneficiaries and subordinates, to use Gilroy's terms, along with the role of genomics in destabilizing racial thinking. To this end, we will examine a case study that explores the prominent role of the new genetics in dealing with a particular set of contemporary sociopolitical issues, notably the relationships between race, biological diversity, and identity construction. We will take as our case study research into the genetic traits of the Brazilian population, based on analyses of mitochondrial DNA, the Y chromosome, and nuclear DNA; we will also look at how this research has been received. The research in question consists of a series of studies that we will call the "Molecular Portrait of Brazil," coordinated by geneticist Sérgio Pena at the Universidade Federal de Minas Gerais (UFMG). The findings have been published in Brazilian popular science magazines like *Ciência Hoje* and in specialized periodicals like the *American Journal of Human Genetics* and *Proceedings of the National Academy of Sciences of the United States of America* (PNAS), as of 2000. Not only have these studies had an impact on academic circles, but they have also received considerable attention in the national and foreign press, fueling heated discussions among specialists and provoking comments by key players in social movements.

The "Molecular Portrait of Brazil" has been received enthusiastically in many circles (see references in Santos and Maio, 2004). Some people consider it a conclusive demonstration of genetics' potential in reconstructing the biological history of the Brazilian people. Journalist Elio Gaspari called the work a "phenomenal article,...a veritable lesson, a source of pride for Brazilian science." He also wrote that "it is the scientific proof of what Gilberto Freyre set out in sociological terms," referring to the magnitude of miscegenation in Brazil. "There are more people [in Brazil] with one foot in the kitchen than with both in the drawing room" (Gaspari, 2000), an expression that was even used by former president Fernando Henrique Cardoso during his mid-1990s campaign.

However, as black rights activist Athayde Motta sees it, this research by geneticists ("using high technology") provides the myth of racial democracy in Brazil with a "simulacrum of scientific support." Further, the findings could open the doors for an "almost infinite potential for manipulation," including the possibility of "bringing new blood to the dying myth of racial democracy" (Motta, 2000a, 2000b; 2002) or even engendering "a pro-racial democracy campaign..., a political and ideological discourse whose prime function is to maintain the state of racial inequalities in Brazil" (Motta, 2003).

Another no less critical assessment of the work by the Brazilian geneticists came from a far right-wing, neo-Nazi group called Legion Europa, based in Europe and the United States. One M. X. Rienzi, author of countless texts on the group's website,[1] wrote that "the authors [the UFMG researchers], in the most shameless, subjective, and unscientific manner, openly display their political views on the subject of race," adding that "it is time to stop to deform natural realities to match political ideologies, and instead accept the racial realities which exist and deal with them as best we can."

As these reactions clearly demonstrate, the work by the Brazilian geneticists has received such attention and had such an impact that at one point there appears to have arisen a curious "proximity" between a black rights activist and a member of a far right-wing group. However different their political leanings and proposals, both criticize the "Molecular Portrait of Brazil," largely accusing it of using science to produce a "ideological and political discourse" whose consequences run counter to their respective worldviews.

Taking these contents and reactions as our backdrop, there are a number of questions we plan to address in this chapter. First, what has been the role of the new genetics in this apparent "proximity," and how did it come about? Also, how are essentialism, racism, racialism, and identity formation expressed in these criticisms? How do politics and science intermingle in these discussions, whose boundaries are clearly not restricted to the physical space of the molecular biology laboratory? Using contrast as an analytical lens, we intend to make a meticulous reading of the positions taken by segments of society that are so apparently disparate in ideological terms, so as better to understand some of the relationships between anthropological and genetic themes in the contemporary world, focusing on the issues of race, race relations, and national and international sociopolitical projects. We are interested in exploring to what extent emerging genetic knowledge may hold sway in influencing and even transforming notions of social coherence and identities, and how organized groups are responding to this.

The Diversity of the Brazilian People from the Perspective of Genetics

In an influential work produced in the 1960s, geneticists Francisco M. Salzano and Newton Freire-Maia stated that the Brazilian people presented "an unrivalled opportunity for the study of some of the most fascinating and complex problems" (1967, p. 1). They noted that "Brazilian populations are generally characterized by great genetic heterogeneity.... The heterogeneity derives from the contribution made by their formative racial groups.... Our populations are thus excellent material for a series of studies on intra- and inter-ethnic comparisons, as well as on the effects of miscegenation" (Salzano and Freire-Maia, 1967, p. 157). In the 1960s and 1970s, many studies were made of Brazil's "racial mix" (see Sans, 2000). They were based on analyses of classic genetic markers, like the rhesus factor, Diego blood group systems, and gamma serum protein (gamma-globulins).

It is within this context of the history of genetics in Brazil that the set of research we call the "Molecular Portrait of Brazil" finds its place. One might say that it is the latest chapter in a major line of investigation into human genetics that flourished in Brazil in the second half of the twentieth century. Yet even more so, the research by Sérgio Pena and his collaborators, along with other genetic studies (see Salzano and Bortolini, 2002; Callegari-Jacques et al., 2003), both innovates and broadens analytical potential by using a new battery of techniques sourced from molecular biology. By sequencing portions of mtDNA and the Y chromosome, geneticists sought to map out a comparative panorama of the geographical distribution and patterns of the Brazilian population's matrilineal and patrilineal ancestry. In line with the considerable literature on the genetics of Brazilian populations (including a current of thought that considers Brazilians "unparalleled and fascinating" due to their high degree of miscegenation), the aim was to unlock the history of the formation of the Brazilian people in biological terms, paying special attention to the social and demographic reality of the country in terms of miscegenation.

The first article from the "Molecular Portrait of Brazil" series came out in Portuguese in 2000 (Pena et al.) in the monthly science magazine *Ciência Hoje*, published by the Sociedade Brasileira para o Progresso da Ciência (SBPC). Two directly related articles containing a detailed presentation of the findings for the scientific community were published in the *American Journal of Human Genetics* (Alves-Silva et al., 2000; Carvalho-Silva et al., 2001) while another came out more recently in the PNAS (Parra et al., 2003).

In the investigation into Y-chromosome DNA polymorphisms, which involved some 250 men from different parts of the country who deemed themselves "white," the overwhelming majority of the markers identified were of European origin, with a very low percentage of sub-Saharan African markers and none that were Amerindian (Carvalho-Silva et al., 2001). Meanwhile, the results of the mitochondrial DNA analyses of the same group indicated a more complex picture, with the sample displaying 33 percent Amerindian and 28 percent African markers, that is, a surprisingly high Amerindian and African matrilineal ancestry among the white Brazilian men under study (Alves-Silva et al., 2000).

According to the authors of the "Molecular Portrait of Brazil," the pattern of differential reproduction detected in the genome analyses (with patrilineal ancestry checked via the Y chromosome, found to be of a predominantly European origin, and matrilineal ancestry checked via the mitochondrial DNA, found to be of an overwhelmingly African and Amerindian origin) makes absolute sense when viewed in the light of the history of colonization in Brazil as of the sixteenth century: "The first Portuguese immigrants did not bring their women, and historical records indicate that they soon began a process of miscegenation with indigenous women. With the arrival of slaves beginning in the latter half of the 16th century, miscegenation extended to African women" (Pena et al., 2000, p. 25). The genetic research findings corroborate the miscegenated nature of the sample of (self-classified) white Brazilians, since the majority (around 60 percent) of the matrilineal lineages were of Amerindian or African origin.

While the two articles published in the *American Journal of Human Genetics* focused primarily on genetic-molecular and phylogeographical aspects, the text written for the purpose of science communication in *Ciência Hoje* made no bones about the social and political implications in the fight against racism in Brazil that could be drawn from the research:

> Brazil is certainly not a "racial democracy."...It might be naïve on our part, but we would like to believe that if the many white Brazilians who have Amerindian and African mitochondrial DNA became aware of this, they would be more inclined to value the exuberant genetic diversity of our people, and perhaps build a more just and harmonious society in the 21st century (Pena et al., 2000, p. 25).[2]

In January 2003, the geneticists published another paper, entitled "Color and genomic ancestry in Brazilians." Unlike the previous studies, which involved individuals from different parts of Brazil, the research by Parra et al. (2003) was carried out in one specific rural community (Queixadinha), in Vale de Jequitinhonha, northern Minas Gerais state. First, a group of approximately 170 people were classed by two researchers as "white," "intermediate," or "black" according to morphological criteria (which the authors called a "clinical assessment"). They took into account features such as skin pigmentation, hair color and texture, and shape of nose and lips. Biological material (blood samples) was then collected from each individual and analyses were performed on a battery of ancestry informative markers (AIMs) gleaned from their nuclear DNA. For comparative purposes, samples were also analyzed from three other groups (Africans from São Tomé, indigenous Amazonian peoples, and Portuguese).

The main finding of the study by Parra et al. (2003) may well have been that there was no direct relationship between morphological and biological classifications in the sample from Queixadinha, in which there was a great deal of overlapping, and it proved hard to distinguish the genomic features of the individuals morphologically classed as "white," "intermediate," and "black." Yet comparison of the genetic characteristics of the other three groups (Africans from São Tomé, indigenous peoples from the Amazon, and Portuguese) showed some marked differences. The authors concluded that: "Our data suggest that in Brazil, at an individual level, color, as determined by physical evaluation, is a poor predictor of genomic African ancestry, estimated by molecular markers" (Parra et al., 2003, p. 177).

Reception and Criticism of the Genome Studies

Athayde Motta and the View of the Black Movement

In a prior work (Santos and Maio, 2004), we described and contextualized black rights activist Athayde Motta's criticisms of the research by Sérgio Pena and his collaborators. We review some of the key points below.

Motta has written at least four texts that are highly critical of the geneticists' work, published in *Afirma: Revista Negra Online* (Motta 2000a, 2000b, 2002, 2003). The key aspects emphasized are: similarities between the "Molecular Portrait of Brazil" and interpretations of Brazilian history, culture, and society that are regarded as erroneous and outdated; a questioning of the importance of genetics in defining collective identities; and the impacts the genetic findings could have on the implementation of public policy designed to fight racism in Brazil.

In his text "Genética para as massas" (Genetics for the masses), Motta (2000a) emphasizes that there are parallels between the interpretations of the geneticists and what he calls other portraits of Brazil's colonial past. The author is referring to the writings of Gilberto Freyre:

> [The "Molecular Portrait of Brazil"] is not far from the colonial portrait of a country initially formed by indigenous populations and white men and later by indigenous and black populations as well as more white men than white women. Considering that it was the Portuguese who had the habit of brutalizing indigenous and black slave women, the research confirms genetically what was already known by anyone with a modicum of critical sense about Brazil.

In his argument, Motta also seeks to undermine the importance of genetic evidence in defining identities and setting patterns of sociability in Brazil:

> The information that 60% of the white Brazilian population is of black and Indian descent might provide some fuel for those who like to say that there are no whites in Brazil, but genetics is not what is going to make this possible. The race and cultural relations in our society are such that a definition of what it is to be white is far from being an issue of genetics or biology (2000a).

Motta's most pointed criticisms of the "Molecular Portrait of Brazil" concern the implications this genetic data may have for public policy. While admitting that "the almost unlimited potential for manipulation…is the fault neither of the research nor of the researchers" (2000a), and that the geneticists' work uses "high technology and good intentions to produce a genetic map of a sample of the white Brazilian population" (2000b), he states that the study provides a "simulacrum of scientific support" for the "myth of racial democracy."

Let us now look into some of the key points of Motta's criticisms. First, they are based on the text published in *Ciência Hoje*—which pays scant attention to technical issues—and contain no reference to those that came out in specialized periodicals (*American Journal of Human Genetics* and *PNAS*). Second, Motta highlights what he considers to be an alliance between the geneticists, on the one hand, and conservative ways of "explaining Brazil," on the other, especially through the ideas of Gilberto Freyre. Third, he questions how relevant biological knowledge is in "revealing" historical and social realities in Brazil (i.e., genetics has demonstrated nothing that history, anthropology, and sociology have not already shown),[3] as well as questioning its role in public policy making.

M. X. Rienzi and the Outlook of the Far Right

Legion Europa is a far-right neo-Nazi group whose website contains information about its convictions and political goals, as well as many texts of a doctrinal nature (see endnote 1). It is impossible to pin them down geographically (i.e., there is no postal address), although from the topics addressed, it can be deduced that they are based in Europe or the United States. The website presents a number of analytical essays about research into human genetics, most authored by M. X. Rienzi (a pseudonym), identified as a biologist from New England. One of these discusses the work of the Brazilian geneticists.

Right at the introduction to Legion Europa comes the answer to a question: "Who Are We?" Their line of argument is that their goal is to reverse the ethnic and sociopolitical weakening of "Euros," which has supposedly been triggered by the influence of other "races" deemed inferior and parasitic. It reads:

> We are Europeans (Euros), or people of various Euro-ethnic descent who partake of a common bioculture/biohistory traditionally known as Western Civilization. Our race is the soil from which this garden bloomed. We are the bees who gather the best of each flower to make the sweetest honey, but we shall also sting mightily people with a common bioculture/biohistory. It is through the betrayal of our people that the OTHER, those not of our race—outgroups—have been enabled to prosper and attain such an influence in world events that our very existence is now threatened by them. ("Other Races")

The website is permeated by a discourse based on assumptions of Aryan racial superiority, militarism (there are numerous references to the Spartans, for instance), anti-Semitism (along with outcries against Arabs and Indians), and a valuing of "German National Socialism" (Hitler Youth and the SS as model corporate structures), among other aspects.

Another feature of Legion Europa's discourse is the degree to which it prizes scientific and technical knowledge in the field of biology and especially genetics. This is what one notes when one reads the "Ethnoracial Bill of Rights" (an obvious parallel with the United Nation's "Declaration of Human Rights"), which contains repeated references to the power of biological technologies to promote the renewal of standards of "homogeneity and coherence" for "Euros":

> Every ethnoracial group, including all peoples of European descent have the right to survive.... Every ethnoracial group has the right to establish whatever degree of biological and cultural ethnoracial homogeneity in the lands in which they live, including the right to establish fully homogenous nation-states, excluding other ethnoracial groups.... The ethnoracial group is an extended kin-group, an extended family. Just as a person must have the right to promote the interests of their family, so must they have the right to promote the interests of their ethnoracial group. There must be no laws that prevent people from fully promoting their ethnoracial interests. There must be full freedom of speech and assembly, full freedom to form political parties, full freedom to promote ethnoracial homogeneity and separatism, and full freedom to oppose globalism, ethnoracially variant immigration, and any other infringement of ethnoracial rights.... Ethnoracial groups must be allowed to pursue whatever reproductive strategies they wish, including endogamy, eugenics, and human cloning.... The pursuit of ethnoracial interests must be accorded the highest priority and dignity. All must be allowed to promote the best for their people, as long as such pursuit does not unfairly infringe upon the ethnoracial rights of other groups.

Rienzi's essay on the research of Brazilian geneticists, published in *PNAS* in 2003, should be read within the context of this extremist racism. Before we look at the criticisms per se, we might attempt to infer what led Legion Europa to take an interest in the Brazilian researchers' work. Aside from the obvious interest of the topic itself (race and the genetics of peoples), the periodical in which the article appeared is considered one of the most influential and prestigious in the world.[4] Published twice monthly, this journal generally considers for publication papers either written or recommended by members of the United States' National

Academy of Sciences.[5] This fact of publication alone would have afforded the text by Parra et al. special visibility, but it was also selected by the PNAS press office to be included in its tip sheet, which is circulated to the press in advance of each new issue so that journalists can prepare their reviews before the journal is published. Of the 40 or so articles included in the issue in question, only three others were selected for inclusion in this press brief, which assured the genetic research ample publicity not only in Brazil but all over the world also.

Rienzi's criticism of the Queixadinha study takes up around four pages; it is lengthy and detailed. Its title is a question: "Scientists Prove Race Does Not Exist?" As the critic from Legion Europa sees it, the work by the Brazilian geneticists is an "ideological tool in the guise of science."

Rienzi starts off his comments by stating that the work by the Brazilian researchers circulated widely across the Internet when it was published. His concern is that people may "believe" that the results of the Brazilian study could be extrapolated into other contexts. To make his point, he mentions a part of the *PNAS* publicity material in which it states that the study shows that no clear correlation is found in Brazil between physical and racial traits and genetic markers of origin and ancestry. As Rienzi sees it, according to those who deny race, the finding by the Brazilians "proves" that the concept of race is unfounded from a biological standpoint. Quoting from an interview given by Sérgio Pena at the time the work was being published, in which he states that the study's conclusions are applicable only to Brazil and "should not be naïvely extrapolated to other countries," Rienzi concludes by stating that the information disseminated to the general public was that the results obtained in the Brazilian study would be universally applicable.

After some rather moderate introductory words, Rienzi's tone turns vitriolic. His comments are even targeted against the epigraph by Parra et al. ("You see leaders today, all over the world, doing it again! Black, white, yellow, brown, people of every color slaughtering people of every color! Because Satan is always the same"; from *The Emperor of Ocean Park*, by S. L. Carter). For Rienzi , this reveals the geneticists' "real" views about the race issue in a "shameless, subjective, and unscientific manner." He asks how a periodical like the *PNAS* could publish a text so imbued with "sentimental subjectivity."

The rest of Rienzi's observations consist of a succession of theoretical and methodological questions that reveal his specialized and thorough reading of the geneticists' text. He goes into detail on the characteristics of the samples, the classification criteria used, the number and type of genetic markers adopted, and the interpretation of the tables, graphs, and statistical tests. He concludes that:

> In summary, this paper does not, in any way, shape, or form, invalidate the biological race concept, and for anyone to make that claim is highly irresponsible to say the least. . . . One wonders if, as regards issues dealing with human biology, we are now dealing with the same kind of entrenched, socio-politically motivated establishment that Galileo dealt with in his work on astronomy. If we let political correctness 'inform' human-genetic scientific work, we are headed back to the days of the Inquisition. . . . It is time to stop attempting to deform natural realities to match political ideologies, and instead accept the racial realities which exist and deal with them as best we can.[6]

A Remote "Proximity": Egalitarian Racialism and Hierarchical Racialism

Paul Brodwin, whom we mentioned in the introduction to this chapter, wrote a comment that somewhat undoes the boundaries between the laboratory and society: "Tracing our ancestry—via a pattern of particular alleles, or mutations on the Y chromosome or in mitochondrial DNA—has become not just a laboratory technique, but a political act" (2002, p. 324).

As he sees it, whatever the answers supplied by genetics, the premises and repercussions are many and significant. Which agents requested the tests, and who provided the samples? Who interprets the findings, and who publishes them? In what contexts are any new interpretations presented to the public? How are they used?

In the sections above, we have reviewed two different kinds of reactions, one by a black rights activist in Brazil and one by a spokesperson from a far-right European-American movement, that focus on the findings of genome investigations carried out recently in Brazil. Both are openly critical of the findings, repercussions, and implications of the research by the geneticists from the UFMG. Athayde Motta refers to the "Molecular Portrait of Brazil" as a "political and ideological discourse whose prime function is to maintain the status quo of racial inequalities in Brazil." M. X. Rienzi, for his part, considers the studies by the Brazilian scholars unscientific and of a political and ideological nature designed to discredit "natural realities," that is, the existence of racial differences and hierarchies.

What we call "proximity" (duly placed between quotation marks for reasons we will explore below) actually evinces a number of things, including the huge influence and power that genetic knowledge wields in the contemporary world, in this case as a source for questioning the notions of identity and the cohesion of social groupings. It has reached such a status and attained such visibility and legitimacy that it ultimately draws into apparent "proximity" dimensions and social players that are actually greatly distanced from each other on any ideological or political plane.

In *On Human Diversity: Nationalism, Racism, and Exoticism in French Thought*, historian and philosopher Tzvetan Todorov introduces a conceptual distinction that may be of use in understanding this "proximity" to which we refer. He underscores the difference between "racialism"—a matter of ideology, a doctrine concerning human races— and "racism"—a matter of behavior, usually involving hatred and disdain for people with clearly defined physical traits that distinguish them from others (1993).

Before we discuss Motta and Rienzi from a racialism-racism perspective, we must first comprehend what stance modern genetics (including current genome research in Brazil along the lines of the "Molecular Portrait of Brazil") takes toward this conceptual duality.

Criticisms of the concept of "race" based on the genetics of populations and on neo-Darwinism have been around for decades. For example, they bore an influence on the first declarations on race drawn up by UNESCO in the 1950s. In the postwar agenda to combat racism around the world, an antiracialist conceptual framework was clearly present. This agenda forwards the notion that the concept of race is not scientifically founded and has little to add to any understanding of human biological diversity. It would be expected that this stance would eventually erode some of the key conceptual building blocks (the existence of races) that have led to discriminatory treatment and the reproduction of social inequality based on race. To a certain extent, this has been the position taken by a large group of geneticists. As the "UNESCO Declaration on Race" shows, a school of biology—or at least a group of researchers—emerged after the war that advocated a "post–World War II universal man, biologically certified for equality and rights to full citizenship" (Haraway, 1989, 1997). Thanks to the efforts of a group of biologists, including Theodosius Dobzhansky and Julian Huxley, it was possible for evolutionary biology and humanism to work hand in hand to reign in aggression and encourage cooperation, dignity, and progress among humankind after World War II (see Santos, 1996; Maio, 1998).

The "Molecular Portrait of Brazil" is a direct descendent of this influential universalist tradition, which is concomitantly antiracialist and antiracist, and which characterized a sizeable portion of the research on the variability of human biology throughout the second half of the twentieth century.[7] Even if its biological bent has not been fully accepted, the interpretative proposition derived from genome research has found a most receptive audience in many circles in Brazil, especially because of its implications. Even as it becomes increasingly clear that Brazil is not a "racial democracy," as socioeconomic statistics show, the country is still seen as racially and culturally hybrid. Dear to broad swathes of the Brazilian society, this stance argues that it is not easy to make out precise compartments, which by and large ends up neutralizing any sharply defined racial identities. As genome research has gained authority and esteem, genetic studies have displayed common points with and provided support for this current, even if geneticists do argue that the concept of race has limited relevance in biological terms. Above all, the narratives about the (bio)history of the formation of the Brazilian peoples that have been produced by genome research along the lines of the "Molecular Portrait of Brazil" are in tune with a deep-seated social imagination that sees miscegenation as a positive, defining element in the identity of Brazil as a nation.

The apparent "proximity" between Motta and Rienzi becomes an infinite distance when one notes that in the former case a racialist yet eminently antiracist set of assumptions predominates, whereas in the latter, an extreme racialism-racism conjugation is what prevails. Yet despite such huge differences, they do hold one point of common ground concerning genome research, which is their criticism of the proposition to dissolve (biological and racial) identities that follows from the "Molecular Portrait of Brazil," that is, its support of an antiracialist viewpoint.

In Motta's view, Brazil displays a system of "archaic and perverse" race relations, which ultimately masks existing discrimination and prejudice and helps assure continued racial inequality. He sees the antiracialism stressed by genetics—as expressed in the "Molecular Portrait of Brazil" and other genetic research—as undermining the cornerstones of the collective identities needed for organizing resistance to oppression. Designed to strengthen racial identity, compartmentalization and polarization are forms of sociability that should be implemented through political actions meant to fight racism, much as what happened in the United States.

If Motta's racialism in theory aims to overcome inequities (his criticism of the work by Brazilian geneticists has much to do with the implications that these genetic data may have in discussions about the introduction of affirmative action policies in Brazil), Legion Europa's racialism has the precise goal of establishing and reinforcing inequities on many levels. In other words, an "egalitarian racialism" predominates in one, while in the other a "hierarchical racialism" prevails. Legion Europa's proposal makes a dangerous link between racialism (with a strong biological tenor) and racism, which Todorov sees as producing especially disastrous results, as was the case with Nazism (1993).

The elements already discussed—compartmentalization, polarization, antagonism, and conflict, again with a view to strengthening racial identities—inform and form the essence of Rienzi's applications of the work by the Brazilian geneticists. Legion Europa's emphasis on compartmentalization is (euphemistically) built on metaphors that play on the words "garden," "flower," and "honey" ("Our race is the soil from which the garden known as 'Western Civilization' bloomed. We are the bees who gather the best of each flower to make the sweetest honey, but we shall also sting mightily those that dare tread on our feet"), under which lies a powerful dose of racial hatred, eugenics, prejudice, and hierarchy, represented by the sting of the bee.

Relativizing Polarities

Almost three decades ago, Claude Lévi-Strauss published his book *View from Afar*, which contains a chapter entitled "Race and Culture." This title echoes a well-known work of his, *Race and History*, which was written on commission by UNESCO during its antiracist drive in the post-Holocaust years (Lévi-Strauss, 1960 [1952]). At one point, Lévi-Strauss refers to the "appearance of the genetics of peoples on the anthropological stage" (1986, p. 14). "Race and culture" shows that the overlapping of anthropology and genetics is not as recent as one might think.

These days, when anthropologists reflect upon and write about genetics, it is not uncommon for them to make frequent references to the notions of "biodeterminism" or "bioreductivism." This is what Roger Lancaster (2003, 2004) suggests about contemporary US anthropologists. As he explains, "Over the past decade, biomythology has permeated American culture as never before. The idea that gender norms, sexual orientations, and social institutions are genetically (or neuro-hormonally) 'hard-wired' flourished in the long shadow of the Human Genome Project" (2004, p. 4). Stressing the role of sociobiology and evolutionary psychology in this process, Lancaster does not attribute the dissemination of bioreductivist viewpoints solely to the expansion of certain fields of science, but above all to the ways in which scientific knowledge is communicated by the media.[8] The media make plenty of room for explanations about how a small set of elements or how this or that gene or biological structure "determine" this or that complex characteristic; they also comment on the development of drugs or other technologies to cure diseases or ame-liorate complex social problems. Lancaster believes that this type of infor-mation is "easy" to convey and readily absorbed by the general public because of its oversimplified cause-effect-solution reasoning. Even if such ideas are refuted by later research, it is unlikely that the disproving of any given bioreductivist formulation will gain as much space in the media as that accorded the original release of information.[9]

Lancaster also notes that certain debates about identity politics in the United States place a sharp emphasis on essentializations, with affini-ties to bioreductivism. Some sectors of the gay movement, for instance, embrace the notion of a "gay gene" to bolster legal arguments pertinent to civil rights discussions. In other words, segments of organized social groups adopt bioreductivism propositions with roots in biology and use these when they devise political tactics for defining and strengthening identities. He comments on this association:

> Identity politics, the quintessentially modern, American justification for social action and political redress by appeal and deep-seated, essential identities, provides fertile ground for bioreductivism, and everybody— the marginal or oppressed and dominant alike—wants to get in on the act.... More than anything, today's reductivism offers to stabilize identity in the *points de capiton* of biology—that is, it purports to secure stability and certitude in an era when nothing much seems anchored about either identity or biology. Furthermore, this approach to securing basic rights and recognition resonates with a long-standing Western understanding of 'nature' as that which exceeds conscious control and volition (2004, p. 5).

How does this digression relate to our case study? In many ways— including the fact that discussions of genome research in Brazil are, at their core, discussions about identity politics.

If, in the case set out by Lancaster, social movements can draw upon biological reductivism, incorporating certain assumptions into their political actions, then in the debates sparked by the "Molecular Portrait of Brazil," what we see is science undermining some of the cornerstones of identity politics. The genetic research conducted in Brazil shows that what we have is not so much profound, immutable essences but rather the "revelation" of a remarkable mixing. With a bit of rhetorical license, one might say that the results of DNA sequencing show that appearances can be deceptive; if we look under their skin, we find that to a greater or lesser extent, whites are genomically "African" and blacks are genomically "European." A subliminal message conveyed by the "Molecular Portrait of Brazil" is that phenotypes and genotypes may be very far removed from each other. These are arguments that rely on emphasizing the fluidity, instability, and ill-defined nature of racial categories.

Whether linked to the black movement like Motta or, above all, to the Far Right like Rienzi, parts of society represented by organized groups view the antiessentialist discourse in the "Molecular Portrait of Brazil" as a "threat" to their basic assumptions, in varying degrees.[10] We would add that the antiessentialist perspective that can be inferred from genome research could come to play a significant role in rhetorical clashes of great social and political import, thanks to the authority and legitimacy that this outlook currently enjoys in Western society.

The discussions kindled by the "Molecular Portrait of Brazil" ultimately rock one of the most prevalent "commonsense" views held by some currents in the social sciences, which see biology (and genetics) as inexorably linked to the proposition and defense of deterministic and essentialist principles. If in Lancaster's examples we see an alliance between a certain line of biological thought and social movements, in "Molecular Portrait of Brazil" a head-to-head battle is waged with science in the defense of an antiessentialism that is considered "threatening" to certain agendas in social and political circles.[11]

Closing Remarks

Throughout this work, we have reflected upon the repercussions of research about the biological and genomic variability of the Brazilian people, and particularly how these studies have fueled clashes and disagreements about assumptions involving extremely broad-based sociopolitical and historical conflicts. In one sense, because of the repercussion of the study published in *PNAS*, Queixadinha—a tiny, poor, rural hamlet in Vale do Jequitinhonha, northern Minas Gerais state, which is rarely even included in national maps, much less in atlases published abroad—has become a playing piece in a game that ultimately is all about discursive conflicts over the ongoing ethnic-racial tension caused by immigration from former African and Asian colonies and Eastern European countries to Western Europe, as well as by the very unification process in Europe. We may well ask ourselves whether, given the emphasis on the genome dimension, Queixadinha might not, under a new guise, represent the idea of Brazil as a prime analytical model in debates on miscegenation, race, and race relations—a role the country has so often played in the past.[12] In this updated version of Brazil as a "paradigmatic" country in which the paradoxes of using the concept of race are exposed by genome research, Queixadinha is seen by the Far Right, represented by Legion Europa, as a hotly contested antimodel.

Our observations throughout this paper also lead us to reflect upon what an "anthropology in the era of genetics" might be. Might we be facing a situation in which new biological technologies directly or indirectly feed the emergence of new ideological configurations? Based on the panoramas we have sketched out, we can state that what we glimpse on the horizon is less a combination of "new biological technologies and new ideological configurations," and more a combination of "new biological technologies and old ideological configurations," to paraphrase Luiz Fernando Duarte.[13] DNA and genomics are enhancing discussions of race, typologies, and nationalisms, against a backdrop of issues in identity and political change that transcend national borders and gain far-reaching international forms.

We can conceptualize the "geneticization" of society as a cluster of changes and a way of generating new meanings within the ambit of Western societies, where the new genetics, or genomics, could be a building block and a key driving force (Lippman, 1991). Commenting on the relationship between geneticization and identities, Paul Brodwin (2002, p. 324) states that "emerging genetic knowledge thus has the potential to transform contemporary notions of social coherence and group identity....What is at stake is also personal esteem and self-worth, group cohesion, access to resources, and the redressing of historical injustice."

As we have argued, the significance of racial differences, and their very essence and existence, are being rebuilt thanks to the impact of genetics. One should ask if this new knowledge and technology will radically alter the scenario or if, on the contrary, they will reinstate and reinforce even more insidious and deterministic ways of perceiving racial differences. In practice, what we perceive is that the relations between biological knowledge and technologies, on the one hand, and racial differences, on the other, may take different forms depending upon their sociopolitical context. We have seen that the so-called geneticization of social dynamics along the lines of the "Molecular Portrait of Brazil" does not necessarily lead to a greater or more ingrained naturalization of racial differences. According to Paul Gilroy, a perspective "against race" is growing out of the research into the genomic make up of Brazilians, which leads to a "deliberate and self-conscious renunciation of 'race' as a means to categorize and divide humanity" (2000, p. 17). Therefore, we must view with a relativizing eye the assumption that a "geneticization" of society, including even its ramifications in the sphere of identity politics, will inevitably go hand in hand with determinism, essentialism, and hierarchy, traits that much of socio-anthropological thinking automatically links to biology.[14]

In the complex, shifting field of interaction between scientific knowledge, racism and racialism, local and transnational contexts, and the agendas of the most varied social movements, the genome-based approach to human biological variability is establishing itself as a tool that can refashion the patterns of proximity and distance between "beneficiaries of racial hierarchy" and "people who have been subordinated by race-thinking," to borrow Gilroy's words. While the ultramodern language of genes and DNA affirms itself as a highly influential element in debates about identity politics in the contemporary world, a perspective on race and essentialized differences endures as an element far from being over-shadowed, but that is undergoing constant reshaping as it interacts with emerging knowledge and technologies.

Notes

First published in *História, Ciências, Saúde-Manguinhos,* Rio de Janeiro, 12(2) (2005), pp. 447–468.

1. The opening page of the Legion Europa website (www.legioneuropa.org) lists the following links: "Who We Are"; "What We Believe"; "What to Do?"; "Euroholidays"; "Ideology of Ethnicity"; "Culture"; "Racial Diversity"; "History"; "Race Reality"; "Commentaries"; "Links." The website was online until at least February 2004, but to date has no longer been accessible (June 2005). All material on the website on November 2, 2003, was printed out, and a copy was filed at the library of the Casa de Oswaldo Cruz, Fundação Oswaldo Cruz, Rio de Janeiro, Brazil.

2. Another example of the potential for linking genetic knowledge and social issues, with implications for public policy making, is the recently published article "Pode a genética definir quem deve se beneficiar das cotas universitárias e demais ações afirmativas?" (Can genetics define who should benefit from university quotas and other affirmative actions?), by Pena and Bortolini (2004).

3. Curiously enough, when the intent of genetic research is to uncover ancestry patterns among black people, Motta paints it in a very positive light. Such is the case of a documentary entitled *Motherland: A Genetic Journey,* produced by the BBC. It contains findings about the genetic origins of Afro-Caribbean Britons. After undergoing genetic testing, program participants traveled to the regions inhabited by their forebears (identified through genomic evidence) so they could "understand a little more about the culture which, to some extent, they share until this day." The narrators add: "It was important for *Motherland* to have the support of a renowned centre of science. Not just to ensure the programme's integrity, but also to cause the scientific world to reflect upon [the fact] that genes and chromosomes may represent much more than defining an animal's gender" (Cesar and Motta, 2004). In Santos and Maio (2004), we comment on this construction of the image of a "genetics for the good."

4. The online version of *PNAS* receives about 4 million hits a week, according to the periodical's website (www.pnas.org/misc/about.shtml, accessed on May 24, 2004).

5. In the case of the article by Parra et al., the recommendation was made by Francisco Mauro Salzano, a leading geneticist who works at Universidade Federal do Rio Grande do Sul (UFRGS) and is one of the only two Brazilian members of the United States' National Academy of Sciences, which publishes *PNAS*.

6. For the purposes of systematization, it is worth highlighting a few aspects of Legion Europa's criticisms. First, they are based solely on the article published in *PNAS* (the papers published in the *American Journal of Human Genetics* and *Ciência Hoje* receive no mention). Second, the criticisms are particularly concerned with how the work by the Brazilian geneticists was widely broadcast by the international media and with the fact that the findings may be extrapolated into contexts other than Brazil. Third, the comments place great emphasis on technical aspects (concerning molecular biology and other methodological questions) and are made by someone who considers himself a member of the same community as the scientists (i.e., the area of genetics of peoples).

7. In Santos and Maio (2004, pp. 86–87), we stated that "within this perspective, the 'Brazilian man' presented by the geneticists, once freed from racist perspectives and aware of his biology, would have a better chance of seeking equality and full citizenship for himself and his peers." In the case of the "Molecular Portrait of Brazil" and other genetic research carried out by Pena and collaborators, the findings that these scientists considered propitious for the strengthening of democratic rights were appropriated or interpreted differently by other groups involved in the debate on race and race relations in Brazil.

8. For more on this, see Horgan (1993), Rose (1997), Condit (1999), and Massarani et al. (2003), as well as many articles in the journal *Public Understanding of Science*.

9. We would like to observe that Lancaster's reflections take on a certain air of nostalgia when he notes that the new bioreductivisms and essentializations currently prevalent in some academic circles and (US) popular culture "not only reverse decades of sophisticated cultural theory and empirical research on cultural variation; they have come to occupy the place once held by anthropology in a progressively dumbed-down serious public sphere" (Lancaster, 2004, p. 4).

10. Athayde Motta had the following to say about the "Molecular Portrait of Brazil": "The information that 60% of Brazil's white population descends from blacks and Indians may provide some fuel for those who like to say there are no whites in Brazil, but it is not genetics that will make this happen. According to our society's patterns of race and cultural relations, the definition of being white is far from a question of genetics or biology" (2000a). For Motta, the heart of the disagreement is not exactly the antiessentialism

of genetics, but an antiessentialism that might spread from its roots in biology to penetrate the social and cultural spheres and become instrumental in defining worldviews.

11. Manuel Castells's perspective on the forms and origins of identity construction is useful for reflecting on our topic. He characterizes a legitimizing identity, a resistance identity, and a project identity. The last of these is present "when social actors, on the basis of whichever cultural materials are available to them, build a new identity that redefines their position in society and, by so doing, seek the transformation of overall social structure" (Castells, 1997, p. 8; also see Calhoun, 1994). The notions of Afro-descendents and Europeans can be understood in the light of the notion of project identity within race relations. What happens is that to a greater or lesser extent genetics, in the form of the "Molecular Portrait of Brazil," destabilizes key assumptions that support these project identities; hence the resistance shown by Motta and Rienzi.

12. As an analytical model for genome studies on race and the biological diversity of the Brazilian people at the beginning of the twenty-first century, it may be that Queixadinha is something of an equivalent to what was represented by research into traditional communities in rural Bahia state, coordinated by Charles Wagley in the 1950s as part of a set of investigations sponsored by UNESCO in the post-Holocaust years. As Wagley puts it in his introduction to *Race and Class in Rural Brazil*, which contains the findings of ethnographic research conducted in various locations around the country, "The world has much to learn from a study of race relations in Brazil.... The various research projects on the subject of race relations which have been stimulated by the UNESCO project in Brazil should give us for the first time an objective knowledge of the situation as it exists under a variety of conditions throughout this vast and variegated country" (Wagley, 1952, pp. 8–9). For more on Brazil in discussions involving race and racism in the postwar years, see Maio (1998; 2001).

13. From a comment made during a discussion of the working group "Pessoa e corpo: Novas tecnologias biológicas e novas configurações ideológicas" ("Person and body: New biological technologies and new ideological configurations"), coordinated by Luiz Fernando Dias Duarte and Jane Russo, XXVII Encontro Anual da Anpocs (Caxambu, Brazil, October 25–27, 2003).

14. For more on this, see the excellent discussion by Peter Wade (2002) in his recent book *Race, Nature and Culture* (especially chapters 5 and 6).

References

Alves-Silva, J. et al. (2000) The ancestry of Brazilian mtDNA lineages. *American Journal of Human Genetics*, 67, pp. 444–461.

Brodwin, P. (2002) Genetics, identity, and the anthropology of essentialism. *Anthropological Quarterly*, 75, pp. 323–330.

Calhoun, G. (1994) Social theory and the politics of identity. In: Calhoun, G. (ed.) *Social Theory and the Politics of Identity.* Oxford: Blackwell, pp. 9–36.

Callegari-Jacques, S. M. et al. (2003) Historical genetics: Spatio-temporal analysis of the formation of the Brazilian population. *American Journal of Human Biology*, 15, pp. 824–834.

Carvalho-Silva, D. R. et al. (2001) The phylogeography of Brazilian Y-chromosome lineages. *American Journal of Human Genetics*, 68, pp. 281–286.

Castells, M. (1997) *The Power of Identity.* Vol. II. *The Information Age: Economy, Society and Culture.* Malden: Blackwell.

Condit, C. M. (1999) *The Meaning of the Gene: Public Debates about Human Heredity.* Madison: University of Wisconsin Press.

Gaspari, E. (2000) O branco tem a marca de Nana. *Folha de São Paulo*, Caderno A, April 16, pp. 14.

Gilroy, P. (2000) *Against Race: Imagining Political Culture beyond the Color Line.* Cambridge: Harvard University Press.

———. (1998) Race ends here. *Ethnic and Racial Studies*, 21, pp. 838–847.

Goodman, A. H. et al. (eds.) (2003) *Genetic Nature/Culture: Anthropology and Science beyond the Two-Culture Divide.* Berkeley: University of California Press.

Haraway, D. (1997) *Modest-Witness, Second-Millennium: Femaleman Meets Oncomouse: Feminism and Technoscience.* New York and London: Routledge.

———. (1989) *Primate Visions: Gender, Race, and Nature in the World of Modern Science.* New York and London: Routledge.

Horgan, J. (1993) Eugenics revisited: Trends in behavioral genetics. *Scientific American*, 268(6), pp. 122–128.

Lancaster, R. N. (2004) The place of anthropology in a public culture reshaped by bioreductivism. *Anthropology News*, 45(3), pp. 4–5.

———. (2003) *The Trouble with Nature: Sex in Science and Popular Culture.* Berkeley: University of California Press.

Lévi-Strauss, C. (1986) *O olhar distanciado.* Lisboa: Editora 70.

———. (1960) *Raça e história* (Raça e Ciência I). São Paulo: Perspectiva.

Lippman, A. (1991) Prenatal genetic testing and screening: Constructing needs and reinforcing inequities. *American Journal of Law and Medicine*, 17, pp. 15–50.

Maio, M. C. (2001) Unesco and the study of race relations in Brazil: Regional or national issue? *Latin American Research Review*, 36, pp. 118–136.

———. (1998) O Brasil no Concerto das Nações: A luta contra o racismo nos primórdios da Unesco. *História, Ciências, Saúde—Manguinhos*, 5, pp. 375–413.

Massarani, L. et al. (2003) Quando a genética vira notícia: Um mapeamento da genética nos jornais diários. *Ciência e Ambiente* (Santa Maria), 26, pp. 141–148.

Parra, F. C. et al. (2003) Color and genomic ancestry in Brazilians. *Proceedings of the National Academy of Sciences of the United States of America*, 100, pp. 177–182.

Pena, S. D. J. and Bortolini, M. C. (2004) Pode a genética definir quem deve se beneficiar das cotas universitárias e demais ações afirmativas? *Estudos Avançados*, 18(50), pp. 31–50.

Pena, S. D. J. et al. (2000) Retrato Molecular do Brasil. *Ciência Hoje*, n. 159, pp. 16–25.

Rabinow, P. (1992) Artificiality and enlightenment: From sociobiology to biosociality. In: Crary, J. and Kwinter, S. (eds.) *Incorporations*. New York: Zone Books, pp. 234–252.

Rose, S. (1997) A perturbadora ascensão do determinismo neurogenético. *Ciência Hoje*, n. 126, pp. 18–27.

Salzano, F. M. and Bortolini, M. C. (2002) *The Evolution and Genetics of Latin American Populations*. Cambridge: Cambridge University Press.

Salzano, F. M. and Freire-Maia, N. (1967) *Populações brasileiras: Aspectos demográficos, genéticos e antropológicos*. São Paulo: Editora Nacional/Editora USP.

Sans, M. (2000) Admixture studies in Latin America: From the 20th to the 21st century. *Human Biology*, 72, pp. 155–177.

Santos, R. V. (1996) Da morfologia às moléculas, de raça à população: Trajetórias conceituais em antropologia física no século XX. In: Maio, M. C. and Santos, R. V. (eds.) *Raça, ciência e sociedade*. Rio de Janeiro: Editora Fiocruz, pp. 125–140.

Santos, R. V. and Maio, M. C. (2004) Qual 'retrato do Brasil'? Raça, biologia, identidades e política na era da genômica. *Mana: Estudos de Antropologia Social*, 10 (1), pp. 61–95.

Todorov, T. (1993) *Nós e os outros. A reflexão francesa sobre a diversidade humana*. Rio de Janeiro: Jorge Zahar.

Wade, P. (2002) *Race, Nature and Culture: An Anthropological Perspective*. London: Pluto Press.

Wagley, C. (ed.) (1952) *Race and Class in Rural Brazil*. Paris: Unesco.

Electronic Sources

Cesar, R. and Motta, A. (2004) Motherland: Uma viagem genética. Available from: http://www.afirma.inf.br [accessed May 26, 2004].

Motta, A. (2003) Contra a genética, o conhecimento. Available from: http://www.afirma.inf.br [accessed February 04, 2003].

———. (2002) Saem as raças, entram os genes. Available from: http://www.afirma.inf.br [accessed September 15, 2002].

———. (2000a) Genética para as massas. Available from: http://www.afirma.inf.br [accessed October 11, 2000].

———. (2000b) Genética para uma nova história. Available from: http://www.afirma.inf.br [accessed October 11, 2000].

2

Molecular Vignettes of the Colombian Nation: The Place(s) of Race and Ethnicity in Networks of Biocapital

Carlos Andrés Barragán

As the celebration of the tenth anniversary of the publication of the first draft of the human genome approaches, the reflections on ethical, legal, and social implications in human genomic research have not decreased but, on the contrary, have become accentuated at the same rate that new breakthroughs in the field of biomedicine get disseminated, and new private and national efforts are mobilized and consolidated to study and account for parts or wholes of populations around the world. The ethical challenges posed by the production, consumption, and governance of immense sets of human genetic data circulating globally are profoundly affected by the emergence of biological capital (Sunder Rajan, 2006), and by the complex political oscillations stimulated by the "scientification" of biomedicine, its socialization, and the "biomedicalization" of society (Burri and Dumit, 2008). On one hand, science-studies scholar Sunder Rajan argues that life sciences are consolidating as one of the main epistemologies of our time, as symptom, a component, or a new phase of capitalism in which life itself has the potential to be valued and dispersed as data (Sunder Rajan, 2006). On the other hand, the contemporary examination of the coupling of life sciences and medicine as culture allows Regula Valérie Burri and Joseph Dumit (2008) to identify the synchronicity of three major processes around health-care innovation and supply during the turn of the century. By "scientification" of biomedicine, the authors mean an increasing shift in the rationalization of medicine that heavily depends on scientific and technological innovation. By contrast, the increasing engagement of different publics with medical knowledge and scientific experts is what they define as "socialization" of biomedicine. Finally, by "biomedicalization" of society, the authors call our attention to the transformations that biomedicine produces in society in general (Burri and Dumit, 2008). These "global assemblages" (Collier and Ong, 2005) or "bioscapes" (Dumit and Burri, 2008) emerging from technoscience, enable the articulation of new grammars through which collective and individual subjectivities, identities, and hierarchies of power-knowledge are being negotiated.

With these perspectives in mind, the production of human genetic and genomic information should be situated beyond the self-referential worlds of scientists and biotechnological agendas in order to establish epistemological dialogues with the experienced worlds that other nonscientific segments of society live in. In this chapter, I present an introductory exploration of the sociopolitical frictions taking place among ethnic minorities and geneticists in Colombia, South America. The arguments are articulated in the form of two vignettes that recount major topics I am finding in my ethnography on the circulation of "ethnic" DNA. In the first one, I focus on some past experiences where the practice of human genetics and genomics has been taking place in the context of highly contested and shifting ethnic politics. In this context, the legacy of previous exchanges between geneticists and ethnic communities—their positive and negative aspects—shape much of the bureaucratic grammars and the generalized resistance that ethnic minority organizations express today about giving new biological tissue for human genomic projects—whether their focus is ancestry studies or biomedicine. In the second vignette, I center my attention on the general frameworks in which "Colombian population" has been understood from an anthropological and genetic perspectives, how certain segments within it have been valued or targeted as a subject of study more than others, and raise the question of why this is changing. In the last section of this chapter, I present some reflections on how, in order to build better and less polarized dialogues among ethnic minorities' organizations and scientists in the case of Colombia, new strategies are fundamental to the engagement of both parties in a mutual process of situated learning of their respective worldviews.

(Not) a Love Affair with Genetics

In 1996, the Colombian Congress hosted an algid debate on the alleged transgression of indigenous people's rights by national and international teams of researchers interested in studying their genetic information. The public audiences consisted of ethnic minority leaders, activists, scientists, and staff of the Ministry of Health and Interior, with the main purpose of evaluating the claims and coming to terms with new forms of governing the access and circulation of human and nonhuman biological materials.

The deliberation was fueled in part by the dissemination of several reports produced by researchers at the Rural Advancement Foundation International (RAFI),[1] an organization dedicated to overseeing international governance of biotechnology. Early in that year, on March 30, RAFI researchers emitted a communiqué titled *New questions about management and exchange of human tissues at NIH: Indigenous person's cells patented*" (RAFI, 1996), which focused on the capitalization and unrestricted circulation of human biological materials between US government institutions and the private industry. The core of RAFI's critique was built around the polemic patent submission made by the US National Institutes of Health (NIH),[2] for an infected cell line extracted from an indigenous Hagahai male in Papua New Guinea (Cunningham, 1998, pp. 210–211); biological material that was collected within the framework of their research on Human T-lymphotropic virus (HTLV) responsible for a type of leukemia and lymphoma. The fact that the NIH failed to provide proper evidence of individual or collective informed consent was considered a red flag for indigenous communities in general, but particularly for those in Colombia; the report read: "Communities whose cell lines are held by NIH have sound reason to be concerned that their cells may be patented" (RAFI, 1996, p. 4; see also RAFI, 1995). RAFI researchers reported that the NIH hold an estimated 2,305 blood samples representing several indigenous, Afro-Colombian, and mestizo communities.[3] They also stated that the US Centers for Disease Control and Prevention (CDC) had unspecified biological materials from indigenous and Afro-Colombian communities, raising a general concern about the purpose of its storage.[4] The samples, according to the NIH records, were collected between 1987 and 1992 and transferred to the NIH through interinstitutional collaborations with the Human Genetics Institute (Instituto de Genética Humana, IGH) at the Pontificia Universidad Javeriana—a Jesuit Order funded, private university in Bogotá D.C. Taking into consideration that at the time some of the research interests on HTLV were shared by both researcher teams at the NIH and the IGH, and most importantly, the rationale behind NIH's previous attempt to patent the Hagahai cell line, RAFI's team stated that "[the] situation signals the possibility that NIH will try to patent Colombian cell line(s)" (RAFI, 1996, p. 6; Cunningham, 1998, pp. 210–211). Indeed, research results on the prevalence of HTLV in the Wayúu community (on the north coast of Colombia) were already published and coauthored by both research teams at the NIH in Bethesda and the IGH in Bogotá D.C. (see Dueñas-Barajas et al., 1992), but without any known attempt to patent its outcome.

RAFI's document was used by ethnic minorities organizations and different NGOs to back up long-term suspicions about the obtention and circulation of human biological materials without the collective consent of the communities. But this was not the first public ventilation of presumed unethical collection of biological tissue among indigenous communities in Colombia.

A year before, in 1995, the British broadcaster *Channel Four* aired a controversial documentary whose title paraphrased Calestous Juma's book title, "*The Gene Hunters*," one of the earliest accounts of the sociopolitical inequalities in the governance of biotechnology between developed and developing countries (Juma, 1989). In the cinematic version of genetic warfare, British director Ian Taylor (1995) presented a story about the collection of human tissue from isolated and "unmixed" indigenous tribes in danger of extinction for the Human Genome Diversity Project (HGDP).[5] Filmed in Colombia and the United States, the documentary followed the fieldwork experiences of a Colombian research team interested in studying the genetic makeup from the Arhuaco and Arsario indigenous communities—both located in the Sierra Nevada de Santa Marta—and the Wayúu indigenous community in the Guajira. In a script comprising fragments of interviews with indigenous spokespersons from Colombia and the United States, Colombian and American geneticists, social scientists, lawyers, and activists, Taylor articulated a narrative of risk and vulnerability for the indigenous communities amid a thick, utilitarian matrix of scientific and corporate interests surrounding the study of human diversity.

The documentary and booklet distributed by *Channel Four* (Taylor, 1995; Wilkie, 1995) rapidly traveled outside the United Kingdom and captured media attention in Colombia. The Colombian scientists framed in Taylor's exposé were biologist and microbiologist Alberto Gómez Gutiérrez and Dr. Ignacio Briceño Balcázar, at the time participants in one of the most valuable and prolific, large scale, scientific research projects carried out in Colombia in the second half of the twentieth century: The Great Human Expedition (*La Gran Expedición Humana*). The project started in 1983, and its initial goal was to explore the Colombian territory in search of human variation, pathologic profiles, and socioeconomic conditions of the population with a multidisciplinary approach—a methodology that was quite novel and ambitious at the time. Conceived within the IGH and directed by Dr. Jaime Eduardo Bernal Villegas in its initial phase the project was supported mainly by the Pontificia Universidad Javeriana during its first years and gradually received support from the government and the private sector (Gómez, Briceño, and Bernal, 2007). The samples and data accumulated during fieldwork over the course of more than ten years, plus the systematic laboratory analysis initiated in 1993, helped make the biological bank at the IGH one of the biggest in the region (Gómez Gutiérrez, 1994; Ordóñez, Durán, and Bernal, 2006).

Initially, Colombian media echoed the claims made in Taylor's documentary (see Castro, 1997a) and the counterclaims by the staff of the IGH (El Tiempo, 1997), until the Colombian researchers finally were able to prove wrong many of the allegations on gene patenting and the connections between the IGH researchers and pharmaceutical companies in California such as Roche Molecular Systems Inc. (at the time a subsidiary of Hoffman-La Roche according to Taylor). Colombian journalist Germán Castro Caycedo rectified his own criticisms against the IGH's scientists when his owns inquiries confirmed manipulation and omission by the director and the producers, for example, about the relation and employment status of the supposedly Roche Molecular Systems researchers that joined the IGH's visit to the Arsario community (Castro, 1997b). Officials from Hoffman-La Roche denied the storage of cell lines from Colombian indigenous communities; also the American researchers were identified as graduate students that indeed did internships at La Roche, but were hired by the producers to portray the interest of big pharma in the immortalization and commercialization of indigenous cell lines (see Castro, 1997b, p. 19). The documentary then helped mostly to create an atmosphere of uncertainty and animosity between the ethnic minorities, human rights activists, and the scientific teams at the IGH. The distrust was also fueled by RAFI's communiqués.

The resulting public discussions held at the Congress in 1996 were themselves an interesting space for the discussion and negotiation of postcolonial structural hierarchies of cultural difference and governance of property and technoscience. The leading indigenous voice in the Congress debates was Lorenzo Muelas Hurtado, the first indigenous leader to reach a position of significant political power within the governmental structure of the country—first as member of the National Constitutional Assembly that produced the new political constitution in 1991 and second, as senator of the Republic in representation of the Special Indigenous Circumscription (1994–1998). Muelas, representing the National Indigenous Organization of Colombia (*Organización Nacional Indígena de Colombia*, ONIC), addressed in his speeches ethnic minorities' collective sense of powerlessness toward the particular style of predatory research practices behind the formation of human genetic banks, the extraction of indigenous traditional knowledge, and the privatization and commercialization of biological diversity. He argued that these processes resembled the colonial order imposed over indigenous peoples since the sixteenth century. Muelas's goal was to elaborate on a proposed law able to encompass the epistemological understanding of indigenous perspectives on biodiversity and its preservation. He wanted to fill the gaps within the legal system pertaining to indigenous autonomy over the governance of biological materials within their communities and territories and protect their human rights within the context of biomedical research.

From a legal perspective, the Colombian Political Constitution (1991) addresses the topic of genetic manipulation in a very vague way, but clear enough to emphasize the sovereignty of the state in the governance of genetic information. In the chapter concerned with collective rights and the environment, the constitution reads that "the State will regulate the entrance and exit of genetic resources, its use, in conformity with the national interest" (República de Colombia, 1991: Chapter 3, Article 81). The discussions in the Congress converged on the necessity of providing the legal mechanisms to protect both ethnic minorities' autonomy and the freedom to carry out scientific research. However, they did not materialize immediately into tangible actions beyond different propositions for the return and repatriation of the biological samples to the indigenous communities (RAFI, 1997)—actions that, at this point in time, have not yet taken place.

A few years later, the topic of human genetic manipulation was officially addressed in Law 599 (República de Colombia, 2000). The law, part of the Colombian Penal Code, sanctioned that any act of manipulation of human genes without free and informed consent would be subject to a sentence fluctuating between one to five years in prison, but did not address what would happen if the biological samples were previously stored or transferred to another country. The general character of this legal measure did not reach the level of detail expected by the ONIC and the Organization of the Indigenous Peoples of the Colombian Amazon (*Organización de Pueblos Indígenas de la Amazonia Colombiana*, OPIAC) leaders. Different proposed laws to regulate genetic manipulation were submitted for deliberation in the Senate in 2001,[6] 2004,[7] and 2005,[8] but all of them were archived whether for incompatibility with other legal dispositions, the lack of unanimity during the voting, and/ or the alleged intervention of high powers such as some Orders of the Catholic Church.

The shortcomings emerging from the negotiations between ethnic minorities and the Colombian government since 1996 have contributed to strong skepticism and increasingly polarized positions about the nature and purpose not only of human genetic research but also any kind of biotechnological research involving animals, plants, and/ or traditional knowledge.[9] During the 1996 debates as well as in other multiple international instances, like the meetings of the Convention on Biological Diversity (CBD), Muelas has defined traditional knowledge as a process that has taken place through the interaction with nature during millenniums and the result of innovation and accumulation of several technologies—knowledge that is transmitted orally from one generation to the next. He points out the existence of biodiversity regions as the most obvious evidence of coproduction. He says, "Biodiversity [nature] is NOT wild" (Muelas, 1998, p. 172).[10] The criticism to Western definitions of traditional knowledge lies in the failure to recognize that such knowledge does not have a separate existence from nature, but is a constituent of it and its transformation, and therefore overflows the dichotomy between tangible (i.e., organic, mineral resources) and intangible (i.e., traditional knowledge). That coproduction between nature and indigenous societies, noted by Muelas, is being mediated by a particular hybrid kinship, where the Earth is considered the Mother, and any exploitation or attempt to profit from it transgresses a subtle equilibrium (e.g., through bioprospecting, patents, commercialization, and privatization).

The Indigenous Authorities of Colombia (*Movimiento Autoridades Indígenas de Colombia*, AICO) rejects the privatization of biodiversity because it cannot belong to anyone in particular. This is a key reflection of their political stand against national and international "sustainable development" discourses and programs in several biodiversity "hot spots" in Colombia, most of them inhabited by ethnic minorities (Escobar, 1998; Ulloa, 2005). In several governmental and nongovernmental instances, Muelas has highlighted that it is impossible not to acknowledge the failure of Western discourse on sustainable development through the consideration of multiple environmental catastrophes across the Colombian territory, as a result of coal and gold mining, oil exploration, and lately the exploitation of biological resources. For Westerners, says Muelas, "what is theirs is theirs; ours, belongs to everyone, to 'humanity,' is theirs" (Muelas, 1998, p. 179). Ethnic minorities in Colombia are skeptical about the effectiveness of multiple local, national, and international mechanisms designed so far for the so-called protection of biodiversity and against biopiracy (Shiva, 1997; Mgbeoji, 2006). From the perspective of the Indigenous Authorities Movement, one of the problems is that these mechanisms look for protection through the immortalization of life in biobanks, while neglecting the racialized genealogies of power that have shaped the indigenous relations with Western society. Muelas framed these relationships in these sharp terms: "[Whites] hate us so much, that they are killing us [ethnic minorities]; but, they love us so much, that they are immortalizing us" (Muelas, 1998, p. 176).

The political intervention that the indigenous collective conceives of is a call for an interethnic moratorium act over any scientific research involving their communities and territories until new regulations concerning informed consent, intellectual property, and democratic participation are negotiated (Muelas, 1998, p. 180). Their strategy embodies a profound epistemological criticism of Western knowledge and the separation between nature and culture, and between human and non-human (Haraway, 1989, 1997); their perspectives challenge and disrupt the discourses of modernity, practicality, development, and innovation that nurture the training and career building process of scientists in Colombia and elsewhere. In the country, the articulation of indigenous identity and social movements—as it emerges in the discursive mode of "ecological natives"—with local and global environmental agendas is what Colombian anthropologist Astrid Ulloa has conceptualized as "eco-governmentality" (Ulloa, 2005); a process that ultimately shapes the medium and discourses of cultural and political autonomy for some ethnic minorities in their attempt to govern biological diversity.

For human geneticists, the near impossibility of accessing new biological tissue from ethnic minorities has boosted, on one hand, the value over the materials stored in genetic banks in different research facilities in Colombia and abroad. On the other hand, I would suggest that this pressure has motivated a reevaluation of the potential that other "admixed populations" such as the mestizo might have for human genetic and genomic research—populations that in previous large-scale research agendas had been in the background for not meeting the criteria of sufficient genetic isolation or purity.

Populations and Identities in the Making

> Race is a fracturing trauma
> in the body politic of the nation
> —and in the mortal bodies of its people.
>
> (*Haraway, 1995, p. 321*)

A reflection on the intrinsic value that ethnic minorities in Latin America had for genetic scientists and doctors during the past decades and in the present cannot evade the consideration and resignification of long-term Western colonial and postcolonial hierarchies of power-knowledge and social representation that have governed the human body and human difference (Haraway, 1989, 1991; Stoler, 1995, 2002; Hall, 1997). In the Amazonian region, for example, the long-term participation of the Yanomami community as the subject of study of anthropologists and biologists and the aftermath that followed the publication of "Darkness in El Dorado: How scientists and journalists devastated the Amazon" (Tierney, 2000), stands as a quintessential case to address the complex articulation of political powers and scientific knowledge when they cross cultural identity and racial and ethnic representation (Fischer, 2001; Borofsky, 2005; Ramos, 2006). Ricardo Ventura Santos has shown how the participation of Brazilian indigenous communities in different genetic research projects, despite the differences in the research styles, agendas, and political contexts, has been framed as key to understanding the biological history of the human species (Santos, 2002). The tensions emerging from the articulation of universal, globalized scientific narratives of humanity with the specificity of indigenous populations who are in turn held up as bearers of attributes of both purity and isolation, translate into particular bioethical challenges both for scientists and communities (Santos and Maio, 2004, 2005; Salzano and Hurtado, 2004; Santos 2006). The experiences of ethnic minorities in the Colombian territory follow similar trajectories and offer interesting insights to reflect on the epistemological constitution of populations from a molecular perspective.

Before considering situated conceptions about cultural and biological difference in Colombia as they emerge within population genetics research, let me contextualize some of the changes within the biopolitical inner workings taking place in the country at the end of the twentieth century. At the end of the 1980s, the Colombian government took a multicultural turn, in part as a response to the countless mobilizations and increasing political pressure from indigenous organizations to achieve political recognition. The new political constitution approved in 1991 declared, as part of its fundamental principles, that the "state recognizes and protects ethnic and cultural diversity of the Colombian nation" (República de Colombia, 1991: Article 7). It also stated that despite Spanish being the official language of Colombia, "ethnic groups languages and dialects are also official within their territories," and that "schooling in communities with their own linguistic traditional would be bilingual" (República de Colombia, 1991: Article 7; 1993). The official recognition of the multiethnic and multicultural character of the nation was without any doubt an important legal and representational step for minorities to strengthen their agendas.

However, despite the discursive progressiveness of the multicultural turn (Pineda Camacho, 1997, 2001; Pardo, 2004), its implementation has not delivered the democratic and pluralist promises or stopped armed

conflict from endangering the well being of all the segments that constitute, paraphrasing Benedict Anderson ([1983]1991), the "imagined" multicultural population of Colombia. In a similar way, the political autonomy
granted by the constitution to ethnic minorities has not changed the
colonial echoes present in the right to define what is, or to decide who
gets to be included as, "ethnic minority" and benefit from the legal rights
that come with that status such as land property (Ng'weno, 2007a). To
date, this imbalance of power acts as the epistemological basis for the
contradictions that emerge when trying to find a clear border between
"national interest" and "ethnic autonomy," particularly in the governance
of biological diversity (see Escobar, 1998; Pardo and Escobar, 2004; Ulloa,
2005). Officially, the Colombian National Administrative Department of
Statistics (*Departamento Administrativo Nacional de Estadística*, DANE),
through the 2005 national census, described the country's population
as being formed by 87 indigenous ethnic groups, 3 differentiated Afro-
Colombian populations and the Rom (Gypsy) population; the rest of the
population gets labeled as "Westerners," the aggregate of whites and mestizos (DANE, 2007).[11]

In Colombia, the official turn toward multiculturalism counterpoints
with unfinished nation-building projects such as *hispanidad* o *mestizaje*
and with a complex hierarchy of differentiation based in the long-term
articulation of racial and geographical identities (Gutiérrez de Pineda
1963, 1968; Guhl, 1965; Wade 1993; Gutiérrez de Pineda and Pineda
Giraldo, 1999) and racism itself (Pineda Camacho, 1984; Wade 1993). A
widespread contemporary narrative about human diversity in Colombia,
transecting school textbooks, official websites, and travel guides for tour-
ists, frames its population in general terms as the historic coexistence and
biological mixture of three primary, pure groups: indigenous natives,
whites, and blacks (see Wade, 1993, 1997); most of the time, each of these
groups are portrayed and circumscribed as biologically pure and historically contained. Efforts to embrace human difference and lessen racism
in Colombia—the multicultural turn as one of them—have introduced
and institutionalized new discursive grammars of difference around indigeneity (Ng'weno, 2007b) in order to define legal ethnic status of indigenous and African descendants' communities (Pineda Camacho, 1995;
Restrepo, 2002). Within this context, legal and commonsensical visions
of race and ethnicity compete and overlap in the definition of cultural
identity when linking ancestry and color (race) and/or ancestry, custom-
ary law, and land rights (ethnicity).

From the three populations legally defined as "ethnic" in Colombia,
indigenous communities have captured with more pervasiveness the
attention of scientific enterprises led by anthropologists and geneticists

during the twentieth century, moved in part by a sense of urgency over the accelerated and irreparable effects that colonization and "acculturation" were having on indigenous people (Barragán, forthcoming). During the 1940s, amid a very racist political understanding of human difference, a first wave of Colombian ethnologists trained at the National Institute of Ethnology (*Instituto Etnológico Nacional*, IEN) and interested in cultural and biological issues carried out several expeditions throughout the "inhospitable" Colombian geography in search of the descendants of "indigenous tribes" mentioned in Spanish Conquest chronicles and for those still "uncontacted." The research agenda influenced and led momentarily by French ethnologist Paul Rivet (1876–1958), founder of the Musée de l'Homme in Paris, had as primary goal the collection of data to reconstruct and support his theories about the peopling process of the American continent and trace the epicenters for the diffusion of cultural and linguistic specificities (Rivet, 1943). The ethnographic expeditions carried out by national and international ethnologists (see figure 2.1) affiliated with the IEN produced the first studies on blood types and physical anthropology of several indigenous communities, both as a scientific enterprise and as an attempt to bring the government's attention to the challenges faced by them due to the colonization of their territories (Barragán, forthcoming).[12]

A second wave of "rescue ethnology" was promoted at the end of the 1960s following the call made by the International Committee on Urgent Anthropological and Ethnological Research, under the direction of Austrian ethnologist Robert von Heine Geldern (1885–1968), to investigate the culture and language of multiple tribes around the world at high risk of being "absorbed into the neighboring populations" and "threatened with physical extermination by epidemics, low fertility, high infant mortality, etc." (von Heine-Geldern, 1959, p. 1076). In Colombia, anthropologist Gerardo Reichel-Dolmatoff (1912–1994) prioritized the ethnographic agenda in order to counteract biological extermination and cultural change taking place due to the spread of government-sponsored colonization (see Reichel-Dolmatoff, 1959; Reichel-Dolmatoff and Dussan, 1966). Regionally, the northwest part of the Colombian Amazon was framed as a priority (Reichel-Dolmatoff, 1967a, 1967b), thus promoting several ethnographic research projects carried out by national and international, undergraduate and graduate students during the next two decades. Reichel-Dolmatoff played a key role in the creation of an intellectual network to foster research activities with indigenous communities in the Amazon region, not only by sociocultural and biological anthropologists, but also by botanists and human geneticists interested in the particularities of indigenous communities and their geographic environments

Figure 2.1 Colombian anthropologist Milciades Chaves Chamorro (1916–1987), preparing to study in situ blood types from members of the Chimila community. Ariguaní River, Magdalena, 1944.

Source: Photographic archive Chaves-Chamorro family.

(Barragán, forthcoming). Some of the most notorious examples of the research networks in human genetics consolidated during this time have been the multiple projects led by American geneticist James V. Neel (1915–2000) and Brazilian geneticist Francisco Salzano on genetic epidemiology since the 1960s (see Salzano, 2000; Santos, 2002); and, since the 1980s, by the Great Human Expedition (Expedición Humana, 2000; Ordóñez, Durán, and Bernal, 2006; Gómez, Briceño, and Bernal, 2007).

A general overview of the intellectual and methodological motivations behind studying indigenous communities in Colombia for research that focuses on human genetic variation, disease gene mapping, and/or genetic ancestry (related to research examining patterns of migration to the American continent and admixture) points out as common denominator their uniqueness as isolated "gene pools." Metaphors used to describe the value that ethnic communities have for geneticists, such as "an uncontaminated paradise" or "Pandora's box" reinforce the speculation about the future potential their genetic information may have for understanding diseases, disorders, and human variation. During the twentieth century the background rationale to study the "enigmatic" biological character of these communities (Neel et al., 1980) was that there is a factor of "purity" that must be studied and preserved before it is too late. An example of how indigenous genetic information is currently valued was offered to me by a researcher[13] working in an academic space and with interests in the legal dimension of the governance of biological diversity, who described ethnic minorities as "cultural property that the Colombian State must protect from its disappearance whether because of the conditions of poverty, endemic givens, or as casualties of the armed conflict." (Anonymous, 2007). Such a claim for protection suggests not only the legitimacy to carry out new research projects but also the possibility of continuing to capitalize on the data already existent in biological banks inside and outside Colombian territory. Such perspective is what motivates, as explored in the first vignette, the very essence of the interethnic moratorium imposed on Western science to access biological sources and traditional knowledge within minority's territories (Acosta, 1994; Muelas, 1998).

The contestation of access to new human tissue from some ethnic communities has produced interesting and challenging frictions that cross multiple social domains. First, it has accentuated the demand over biological materials stored in commercial and noncommercial biobanks; second, it has raised bioethical criticisms and algid debates about intellectual property; and third, it has opened a space for the philosophical and methodological consideration of how populations are being defined and sampled. In the context of research on human genetic variation in

Colombia, tissue samples can "talk" about individuals, communities, groups, ethnicities, races, and populations independently of the time of their collection. When stored samples and processed genetic data is used in comparative research, an interesting suspension and compression of time and space takes place by reifying collective disembodied identities through individual samples that can represent the whole (i.e., the Tukano, the Yanomami, the Kayapó, or more broadly indigenous people in South America, or in Latin America, etc.).

At the discursive level, the reproduction and transmutation of a specific object, the sample, into an individual or collective identity resembles in an interesting way what British ethnologist, Sir James George Frazer (1854–1941), defined as a principle in prescientific thinking: that of "similarity" and that of "contact and contagion." He argued, "If we analyse the principles of thought on which magic is based, they will probably be found to resolve themselves into two: first, that like produces like, or that an effect resembles its cause; and, second, that things which have once been in contact with each other continue to act on each other at a distance after the physical contact has been severed" (Frazer, 1922, p. 11). This insight is particularly useful to understand the contentious imbedded in the governance of "ethnic" biological samples. Secondhand and new samples come to represent individuals and the populations they were part of, allowing scientists to make claims about them. At the same time these claims act upon a new generations of individuals, members of multiple ethnic communities, who may not be interested in engaging with science, or may be offended by the claims made, the global commodification and circulation of their genetic information or remain indifferent to being told "who they are" or "where they came from." This also true for those individuals or communities interested in articulating such knowledge into their political agendas (e.g., racial quotas).

In my research, geneticists are being questioned about the criteria for the selection of individuals to be sampled within specific populations—whether if the individual(s) labeled themselves in terms of race/ethnicity, or if those conducting the sampling process made the assignation. In some cases geneticists' reactions have been to consider that reflection unnecessary: "You need to pay attention to the phenotype (skin color, con-texture) of the individual and the composition of the community where he/she lives" (Anonymous, 2007). When asked about the possible fluidity of ethnic/racial identification that does not necessarily overlap the color line of brown, white, and black phenotype, another geneticist emphasized the necessity of being careful in the selection of the donors by exploring their life histories, but similarly, saw this as unnecessary if you were working with non-Westerners (i.e., indigenous, Afro-descendants, and

Rom communities). Later on, this same researcher added that it wouldn't matter if there were uncertainty on the ethnic background of a person if "you are dealing with *mestizo* populations; but then that would depend on your research questions. It would not matter if you are doing ancestry studies, for example" (Anonymous, 2007).

The easiness with which the fluidity of collective and individual identities can be thought about is quite surprising and seems to be epistemologically less relevant when researchers and technicians are using secondhand human tissue collected by colleagues in the past (that had been already "racially" and/or "ethnically" tagged in some way). The idea of population frequently is taken for granted, as a natural fact, rather than as a constructed, relative concept both for human biology and cultural politics. Similarly, an increasing focus by some scientists on "admixed populations" as a source for ancestry and gene disease mapping purposes (Sans, 2000; Salzano and Bortolini, 2000; Darvasi and Shifman, 2005) is producing interesting resignifications of highly contested racial and ethnic categories such as mestizo and Hispanic when used to define new sample populations and enroll donors that are supposed to represent a very particular process of miscegenation across Latin America, in general (see Wang et al., 2007; Wang et al., 2008; Silva-Zolezzi et al., 2009), and in Colombia, in particular (see Ospina-Duque et al., 2000; Bedoya et al., 2006). What happens when phenotype defies continental, regional, and/or national circumscriptions? What happens when self-ascription does not fit local and global physical trait stereotypes? What are the challenges such considerations present for biomedical assessments, translational genomics, and in general for the compatibility and standardization of genetic databases? How does the relative ease to access tissue from individuals considered in Colombia "nonethnic," or admixed, influence the definition of methodological sampling strategies, and at the same time chart new narratives of ancestry?

The way population genetics and genomic research is aggregating and disaggregating individuals into populations has profound consequences for the ways individuals consume genetic information, how this new knowledge is incorporated into nation-building agendas that celebrate human difference, or how it is used by the scientists themselves to fight racism and geographic segregation present in Colombia (Expedición Humana, 2000; Yunis, 2006). The ethnographic account of how social constructs of human variation informed researchers' perspectives on the susceptibility of individuals on the Mexico/US border for developing diabetes-type 2 is an interesting case to think with. Anthropologist Michael Montoya evidenced that the combination of ethnic attributes and the life conditions of the affected populations ended up "pathologizing" the ethnicity of the DNA donors (Montoya, 2007, 2011); this is what he terms "bio-ethnic conscription," the multistage practice of using racial and ethnic classification to describe human groups and to attribute qualities to them.

The elaboration of human genomic research agendas in Colombia and the argument that this kind of knowledge should be a matter of national concern and a public health priority for the state, does nothing but underline the relevance of opening up the discussion on the new articulations emerging from racial and ethnic identities, science, and social orders. Dr. Jaime Eduardo Bernal Villegas, director of the IGH in Bogotá D.C., argues that in the country genomic research, proteomics, and bioinformatics in general are in a "prehistoric stage." He highlights the existence of independent, valuable contributions, but points out the lack of a large-scale effort in these areas. Since 2006, with the support of the Pontificia Universidad Javeriana, he has been working to create a multidisciplinary research team to explore the relevance these studies have for Colombia in order to start a systematic gene-mapping program (Bernal, 2007, personal communication). The *"Iniciativa Genómica Javeriana"* looks to reach developments in structural, functional, and comparative genomics and biomedicine. From his point of view, Colombia (through the goals already achieved by the Universidad Javeriana and the IGH in the field of medicine and genetics and by other colleagues in other universities in the country) has the potential to make significant contributions in what is already a natural given in the country: biological diversity (Bernal, 2007, personal communication).

The Colombian Genomic Initiative has an holistic and ambitious approach that at the moment lacks the hype of venture capitalism and governmental support behind similar scientific enterprises in countries like Mexico, with its recently created *Instituto de Medicina Genómica*, INMEGEN (see Jiménez et al., 2008), where personalized genomic medicine research is coupled with a strong nationalist accent (Benjamin, 2009). But the road ahead for scientists in Colombia will not be easy considering that recent political mobilizations and legal frameworks may limit the scope of researchers' activities and disrupt the flow of such genetic information. Eventually, these changes could even frame their activities as illegal depending on how property regimes governing biological diversity flow locally and globally (Waldby and Mitchell, 2006). For example, if genetic information is private, individuals can use, profit, and trade information (such as in the case of assisted reproductive technologies). When genetic information is considered collective, as is the case for indigenous and Afro-Colombians communities, it cannot be used, sold for profit, or traded. If genetic information is considered state property, its potential for use, profit, or trade could be different on an individual basis. Lastly, it can also be considered as public, in which case the benefit for the great national majority will define its status (see República de Colombia, 1991: Article 81). The labyrinth of possibilities emerging from this shifting set of regulations—sometimes overlapping, sometimes in contradiction, and more often than not behind the always changing scientific visions, understandings, and technologies that they try to govern—unfortunately has not prevented biopiracy or bioethical transgressions from taking place. Nor has it facilitated less hegemonic and more productive exchanges between ethnic minorities' leaders, donors, and scientists.

Beyond Informed Consent and Genetic Literacy

In the conversations I had with international and national scientists and ethnic minorities leaders in Colombia about their perceptions of each other from previous exchanges (in the field, court, public debates) and about the reasons why some of these experiences went wrong, both parties pointed out lack of understanding and transparency. Further, such issues often get narrowed down specifically to the absence or flawed application of informed consent—understood as a discursive process and as an object, a document. In such context more often than not, it is proposed that informed consent should be enforced more and more in terms of a detailed legal contract.

A geneticist with expertise in biomedicine framed the situation in this way: "We carry a stigma. [Ethical] transgressions constitute a complex antecedent. We were banned in a way, and now we are facing the consequences of having our work and motivations put into question all the time. The only way I envision this situation can be transformed is by doing new research under clear bioethical protocols that specifically address that the samples would not be used for other purposes beyond those told to the communities or individuals" (Anonymous, 2009). From the point of view of some ethnic leaders, such an approach could be helpful for those involved in the dispute of governance and property rights of biological diversity—human and nonhuman (see Greaves, 1994; Harry, Howard, and Shelton, 2000). In the context of Colombia, revisiting legal frameworks and bioethical protocols is a much-needed step (Flórez, 1998). Its urgency is underlined as multiple actors (ethnic leaders, human rights activists, scientists, lawyers, politicians, and intellectuals) focus on and appropriate past and present international experiences involving scientists and minorities when building their political claims.[14]

However, I consider that a commitment to strengthen bioethical protocols, as the main strategy for reaching understanding, does not challenge a top-down discourse that allows geneticists to decide what information ethnic minorities should know or not know according to arbitrary levels of genetic literacy. Concepts such as informed consent are based, as Hilary Cunningham has pointed out, on scripted institutional guidelines— public and private—that cannot address the sociocultural particularities in play because, by default, they take for granted "what 'informed' and what 'consent' mean in a specific context" (Cunningham, 1998, p. 228). The hardening of bioethical protocols on the contrary would prevent biologists, geneticists, and doctors from engaging in ontological oriented *dialogues* with communities' leaders and donors, a most powerful pathway to reach understanding from my perspective. If, on the contrary, such a level of dialogue is pursued, it will contribute to an understanding of how, despite the supposedly universality of Western knowledge (Haraway, 1989; Escobar, 1995, 1998), the language of science will be contested by the local idioms of the communities it is trying to transform into a subject of study (Rapp, 2006). In this context, questions about how much ethnic minorities know about population genetics or human genomics, or how much of their political positioning against it comes from misconceptions or science fiction, lose relevance. Ideally, new questions will become more preeminent, for example, ones that reflect on the discursive and material strategies at stake while both parties talk and mobilize efforts in the governing of "nature" and human and nonhuman biological diversity. As suggested by Sheila Jasanoff and other science studies scholars, a relevant step is to account for the embeddedness between scientific knowledge and social orders, their coproduction (Jasanoff, 2004a, 2004b).

The work ahead for biologists and geneticists practicing in Colombia is enormous considering the fast-paced way in which the life sciences are reshaping individual and collective life experiences and bioeconomic grammars (Haraway, 1997; Sunder Rajan, 2006). Their approach to ethnic minorities needs to be rethought and redesigned beyond short-term exchanges while the collection of samples takes place. The inclusion of communities in the development of the research agendas has proven to be beneficial for facilitating understandings, particularly following the criticisms against the population-sampling strategies adopted by the HGDP (Cunningham, 1998; M'charek, 2005; Reardon, 2005), the Haplotype Map International Project, HapMap (see Reardon, 2007), and National Geographic's Genographic Project (TallBear, 2007). Questions concerning the inclusion and participation of communities, particularly for contemporary genome-wide association studies (Need and Goldstein, 2009), in turn open an interesting opportunity for human genetics researchers to readdress the logics and methodologies behind the selection of samples, the constitution of new populations, and consider the effects that such process have for different segments of society (Reardon, 2007; TallBear, 2007). This is particularly relevant for ethnic minorities and new forms of indigeneity emerging beyond the dualistic race/ethnic framework (Ng'weno, 2007b).

The consolidation of new research agendas in the fields of population genetics and human genomics in Colombia do not necessarily have to end up reproducing colonial hierarchies of power-knowledge while pursuing discursive and material forms of scientific innovation. As a process in the making, however, they have the potential to build upon past experiences to conceive scientific knowledge and practices otherwise, in an attempt to take seriously and "coproduce" with other forms of knowledge and worldviews. But this challenge is not a discovery, and also it is not something that anthropologists "as social experts" can be hired for, in the hope of letting them deal with the "political dimension" outside the laboratories, while scientists focus on data analysis—we (anthropologists) already have a troubled relationship in the coproduction of difference, particularly with minorities. Scientists have witnessed these challenges multiple times not only in the lab, in the field, in the drafting of publications, et cetera, but also, and in particular, when they themselves are situated as living bodies, as patients, as part of a population(s), or racially and ethnically labeled while crossing a national border or receiving a scholarship. The challenge of engaging and entangling through new dialogues can be suggested, again, by emphasizing that the disciplinary territories such as genetics, genomics, epidemiology, anthropology, et cetera, do not only belong to their practitioners but also to many, and for as long as they remain as social domains for contestation (Dumit and Burri, 2008). There is a possibility for all the actors to learn across all the social domains enacting genomic research. In the meantime, racial and ethnic classifications, even through genomic lenses, will keep being "floating signifiers."

Acknowledgments

The arguments presented here are part of the ongoing research project: "Situating genetic expressions: Human genomic research and bio-identity in Amazonia (Colombia and Brazil.)" I want to thank the University of California Pacific Rim Research Program for its generous support to make possible part of my fieldwork. I also want to express my gratitude to Joe Dumit, Sahra Gibbon, Jim Griesemer, Michael Montoya, Bettina Ng'weno, Ben Orlove, and Ricardo Ventura Santos for their ongoing invaluable support with my dissertation research and for the criticisms offered multiple times during the preparation of this chapter. Joanne Goodsell patiently read different drafts; her comments were immensely helpful to reach this final version.

Notes

1. RAFI's main goal since its beginning was overseeing international governance of biotechnology; today, RAFI exists as Action Group on Erosion, Technology and Concentration (ETC Group), based in Ottawa, Canada.
2. The patent was placed by one of NIH's units, the National Institute of Neurological Disorders and Stroke (NINDS).
3. From the 2,305 blood samples, the report reads: "77% (1733) are samples from indigenous people. Of the remainder, about 15% (338) are from Afro-Colombian communities on the Pacific coast and 8% (193) are from *mestizo* (persons with both indigenous and European ancestry) communities." (RAFI, 1996, p. 5; see also RAFI, 1993). Among the indigenous societies represented through the samples located at NIH were: Achagua, Arhuaco, Barasana, Cenu, Chimila, Coreguaje, Cuaiquer, Cubeo, Cuna, Desana, Embera, Guahibo, Guanana, Guane, Guayabero, Kogi, Motilon, Paez, Piapoco, Piratapuyo, Pisamira, Tikuna, Tucano, Tunebo, Waunana, Wayúu, and Wiwa.
4. In 1993, CDC gained international attention and suspicion among indigenous communities for filing a patent application (later withdrawn) on a cell line of an indigenous Guaymi female from Panama that in theory was resistant to the virus that produces leukemia (see Acosta, 1994; RAFI, 1994a; 1994b; Haraway, 1995, p. 354–355; Ramos, 2006).
5. The producers were Luke Holland and Sandra Clark, and the documentary had the financial support of the Ford Foundation and the Television Trust for the Environment (Taylor, 1995; Wilkie, 1995).
6. "Proyecto de Ley 151 de 2001," filed on November 1 and presented to the Senate by Jesús Ángel Carrizoza Franco. The project was archived in conformity with Article 162 of the Political Constitution (Pinedo, 2001).
7. "Proyecto de Ley 2 de 2004," filed on July 20 and presented to the Senate by Bernardo Alejandro Guerra. The project was archived (Moreno de Caro, 2004).
8. "Proyecto de Ley 188 de 2005," filed on November 29 and presented to the Senate by Antonio Navarro Wolff. The project was archived in commission in conformity with Article 190 Ley 5a. (Mesa, 2005).
9. The experience of indigenous communities in Colombia with bioprospecting agendas has not been less controversial. One of the most well-known cases was the patent of *yagé* (*Banisteriopsis caapi, Banisteriopsis* spp.), a vine used in the preparation of several psychoactive drugs. On June 17, 1982, entrepreneur Loren S. Miller obtained from the US Patent Office the patent (PP5,751) on a variety of *Banisteriopsis caapi*. The acknowledgment of this patent by several indigenous authorities led to a formal complaint in 1999 to the now US Patent and Trademark Office (US-PTO) by the coordinator of Indigenous Organizations of the Amazon Basin (*Coordinadora de las Organizaciones Indígenas de la Cuenca Amazónica*, COICA) with the support of the Center for International Environmental Law (CIEL). US-PTO rescinded the patent on the basis that there was a previous description of the invention, but not on the recognition of indigenous' traditional knowledge as invention. Miller

appealed the decision, and in 2001, the US-PTO reinstated the patent although its life span was to expire two years later (June 17, 2003), and under the contemporary legal framework it couldn't be renewed. The *yagé* patent remains as a powerful symbolic loss for Amazonian indigenous societies against the legitimacy of hegemonic understanding of science, knowledge production, and property. It exemplifies what has been denounced as biopiracy—the claim of ownership by individuals or corporations over traditional knowledge, technologies, and genetic resources from societies in the so-called Third World (Shiva, 1997)—and the representational dilemma in its more nuanced, politically correct version: bioprospecting (Hayden, 2003).

10. His emphasis. The word Muelas used in Spanish is *"silvestre,"* to mean that nature is not uncultured (Muelas, 1998).

11. According with the last national census, the Colombian population reached an estimated of 42,888,594 people. Indigenous people registered 1,378,884 individuals (3.4 percent of the total of population). Afro-Colombians registered a total of 4,261,996 (representing 10.5 percent of the total). The total of Rom people was 4,832. See: http://www.dane.gov.co/ accessed on July 2010.

12. The study of the physical characteristics and blood types in these communities was part of a more comprehensive ethnographic research agenda in which the novel ethnologists received training under the general supervision of Rivet (Barragán, forthcoming). Some of the research analyses were published and focused on indigenous communities living in the departments of Caldas (Duque, 1943–1944), Cauca (Lehmann, Duque, and Fornaguera, 1943–1944), Magdalena (Chaves, 1946), Nariño (Lehmann, Ceballos, and Chaves, 1946), Pasto (Páez and Freudenthal, 1943–1944), and Tolima (Reichel-Dolmatoff and Dussan, 1943–1944).

13. Some of the names and the nationalities of individuals whose perspectives are included in this chapter are omitted to protect their identity (in accordance to their particular request).

14. Two cases in the United States have recently received the attention of ethic minorities' organizations, as both get mentioned during my conversations with them. The first one follows the publication of Rebecca Skloot's bestseller on the replication and commercialization of cell lines coming from Henrietta Lacks (better known as HeLa cells), an African American who suffered and died from cervical cancer (Skloot, 2010; see also Landecker, 2000). The second one is the settlement reached after the lawsuit that the Havasupai Indian community filed against Arizona State University for the unauthorized use of tissue for purposes different than ones agreed to when collected (see Harmon, 2010).

References

Acosta, I. (1994) The Guaymi patent claim. In: van der Vlist, L. (ed.), *Voices of the Earth: Indigenous Peoples, New Partners and the Right to Self-Determination in Practice*. Amsterdam: NCIV / International Books, pp. 44–51.

Anderson, B. ([1983]1991) *Imagined Communities: Reflections on the Origin and Spread of Nationalism*. London: Verso.

Barragán, C. A. (Forthcoming) *Antropología en Colombia: 1890–2000*. Bogotá D.C.: Manuscript in Preparation.

Bedoya, G. et al. (2006) Admixture dynamics in Hispanics: A shift in the nuclear genetic ancestry of a South American population isolate. *Proceedings of the National Academy of Sciences of the United States of America*, 103(19), pp. 7234–7239.

Benjamin, R. (2009) A lab of their own: Genomic sovereignty as postcolonial science policy. *Policy and Science*, 28(4), pp. 341–355.

Bernal Villegas, J. (2007) Conversation, August 23, Bogotá D.C.: Instituto de Genética Humana, Pontificia Universidad Javeriana.

Borofsky, R. (2005) *Yanomami: The Fierce Controversy and What We Can Learn from It*. Berkeley and Los Angeles: California University Press.

Burri, R. V. and Dumit, J. (2008) Introduction. In: Burri, R. V. and Dumit, J. (eds.), *Biomedicine as Culture: Instrumental Practices, Technoscientific Knowledge, and New Modes of Life*. New York and London: Routledge, pp. 1–14.

Castro Caycedo, G. (1997a) Tráfico de genes. *Cambio*, 187 (Enero 13), p. 24.

———. (1997b) Un periodismo depravado. *Cambio*, 190 (Febrero 3), p. 19.

Chaves Chamorro, M. (1946) Contribución a la antropología física de los chimila. *Boletín de Arqueología*, 2(2), pp. 157–177.

Collier, S. J. and Ong, A. (2005) Global assemblages, anthropological problems. In: Ong, A. and Collier, S. J. (eds.), *Global Assemblages: Technology, Politics, and Ethics as Anthropological Problems*. Malden and Oxford: Blackwell Publishing, pp. 3–21.

Cunningham, H. (1998) Colonial encounters in postcolonial contexts: Patenting indigenous DNA and the Human Genome Diversity Project. *Critique of Anthropology*, 18(2), pp. 205–233.

DANE (2007) *Colombia una nación multicultural. Su diversidad étnica*. Bogotá D.C.: Dirección de Censos y Demografía, Departamento Administrativo Nacional de Estadísticas, DANE.

Darvasi, A. and Shifman, S. (2005) The beauty of admixture. *Nature Genetics*, 37(2), pp. 118–119.

Dueñas-Barajas, E. et al. (1992) Coexistence of human T-lymphotropic virus types I and II among the Wayúu Indians from the Guajira Region of Colombia. *AIDS Research Human Retroviruses*, 8(11), pp. 1851–1855.

Dumit, J. and Burri, R. V. (2008) Epilogue. Indeterminate lives, demands, relations: Emergent bioscapes. In: Burri, R. V. and Dumit, J. (eds.), *Biomedicine as Culture: Instrumental Practices, Technoscientific Knowledge, and New Modes of Life*. New York and London: Routledge, pp. 223–228.

Duque Gómez, L. (1943–1944) Grupos sanguíneos entre los indígenas del departamento de Caldas. *Revista del Instituto Etnológico Nacional*, 1(2), pp. 624–653.

El Tiempo (1997) No patentamos genes: Universidad Javeriana. *El Tiempo* (13 de Octubre). Bogotá D.C.

Escobar, A. (1995) *Encountering Development: The Making and Unmaking of the Third World*. Princeton and Oxford: Princeton University Press.

———. (1998) Whose knowledge, whose nature? Biodiversity, conservation and the political ecology of social movements. *Journal of Political Ecology,* 5, pp. 55–82.

Expedición Humana (2000) *Geografía humana de Colombia: Variación biológica y cultural en Colombia (Tomo I).* Bogotá D.C.: Instituto Colombiano de Cultura Hispánica.

Fischer, M. J. (2001) In the science zone. The Yanomami and the fight for representation. *Anthropology Today,* 17(4), pp. 9–14; 17(5), pp. 16–19.

Flórez, M. (1998) Regulaciones, espacios, actores y dilemas en el tratamiento de la diversidad biológica y cultural. In: Flórez, A. M. (ed.), *Diversidad biológica y cultural: retos y propuestas desde América Latina.* Santa Fé de Bogotá D.C.: Grupo Ad Hoc Diversidad Biológica / Instituto Latinoamericano de Servicios Legales Alternativos (ILSA)—Grupo Semillas / Instituto de Gestión Ambiental (IGEA) / Proyecto Implementación Convenio sobre Diversidad Biológica (WWF), pp. 29–44.

Frazer, J. G. (1922) *The Golden Bough: a Study in Magic and Religion.* New York: The Macmillan Company.

Gómez Gutiérrez, A. (1994) El banco biológico humano y la paradoja de la conservación de los grupos étnicos minoritarios. *América Negra,* 8, pp. 149-156.

Gómez Gutiérrez, A.; Briceño Balcázar, I.; and Bernal Villegas, J. E. (2007) *Hereditas, diversitas et variatio: Aproximación a la historia de la genética humana en Colombia.* Bogotá D.C.: Academia Nacional de Medicina / Instituto de Genética Humana, Pontificia Universidad Javeriana.

Greaves, T. (ed.) (1994) *Intellectual Property Rights for Indigenous Peoples: a Sourcebook.* Oklahoma City: Society for Applied Anthropology.

Guhl, E. (1965) *Geografía humana de Colombia: Los fundamentos geográficos y los problemas económicos, sociales y culturales, y el hombre en Colombia.* Serie Latinoamericana. Bogotá D.C.: Facultad de Sociología, Universidad Nacional de Colombia.

Gutiérrez de Pineda, V. (1963) *La familia en Colombia: Trasfondo histórico.* Serie Latinoamericana. Bogotá D.C.: Facultad de Sociología, Universidad Nacional de Colombia.

———. (1968) *Familia y cultura en Colombia: Tipologías, funciones y dinámica de la familia, manifestaciones múltiples a través del mosaico cultural y sus estructuras sociales.* Bogotá D.C.: Facultad de Sociología, Universidad Nacional de Colombia / Tercer Mundo Editores.

Gutiérrez de Pineda, V. and Pineda Giraldo, R. (1999) *Miscegenación y cultura en la Colombia colonial: 1750–1810.* Santa Fé de Bogotá D.C.: Universidad de Los Andes / Colciencias.

Hall, S. (ed.) (1997) *Representation: Cultural Representations and Signifying Practices.* London: Sage Publications.

Haraway, D. J. (1989) *Primate Visions: Gender, Race, and Nature in the World of Modern Science.* New York and London: Routledge.

———. (1991) *Simians, Cyborgs, and Women: The Reinvention of Nature.* New York and London: Routledge.

———. (1995) Race: Universal donors in a vampire culture: It's all in the family. Biological kinship categories in the twentieth-century United States. In: Cronon, W. (ed.), *Uncommon Ground: Toward Reinventing Nature*. New York: W.W. Norton & Co., pp. 321–366, 531–536.

———. (1997) *Modest Witness@Second Millennium. FemaleMan©_Meets_ OncoMouse™: Feminism and Technoscience*. New York and London: Routledge.

Harmon, A. (2010) Indian tribe wins fight to limit research of its DNA. *New York Times*, Thursday, April 2, Section 1, pp. A1, A17.

Harry, D.; Howard, S.; and Shelton, B. L. (2000) *Indigenous People, Genes and Genetics: What Indigenous People Should Know about Biocolonialism*. Wadsworth: Indigenous Peoples Council on Biocolonialism.

Hayden, C. (2003) *When Nature Goes Public: The Making and Unmaking of Bioprospecting in Mexico*. Princeton and Oxford: Princeton University Press.

Jasanoff, S. (ed.) (2004a) *States of Knowledge: The Co-Production of Science and Social Order*. New York and London: Routledge.

———. (2004b) The idiom of co-production. In: Jasanoff, S. (ed.), *States of Knowledge: The Co-Production of Science and Social Order*. New York and London: Routledge, pp. 1–14.

Jiménez-Sánchez, G. et al. (2008) Genomic medicine in Mexico: Initial steps and the road ahead. *Genome Research*, 18, pp. 1191–1198.

Juma, C. (1989) *The Gene Hunters: Biotechnology and the Scramble for Seeds*. London: Zed Books.

Landecker, H. L. (2000) Immortality, in vitro: A history of the HeLa cell line. In: Browdin, P. (ed.), *Biotechnology and Culture: Bodies, Anxieties, Ethics*. Bloomington: Indiana University Press, pp. 53–74.

Lehmann, H.; Ceballos Araujo, A.; and Chaves Chamorro, M. (1946) Grupos sanguíneos entre los indios "kwiker." *Boletín de Arqueología*, 2(3), pp. 227–230.

Lehmann, H.; Duque Gómez, L.; and Fornaguera, M. (1943–1944) Grupos sanguíneos entre los indios guambiano-kokonuko. *Revista del Instituto Etnológico Nacional*, 1(1), pp. 197–208.

M'charek, A. (2005) *The Human Genome Diversity Project: An Ethnography of Scientific Practice*. Cambridge and New York: Cambridge University Press.

Mesa Betancur, J. I. (2005) Proyecto de Ley 188 de 2005 por medio del cual se establece la huella genética como medio de identificación y se dictan otras disposiciones correspondientes para la generación de bancos de datos genéticos. *Gazeta del Congreso* 847. Bogotá D.C.

Mgbeoji, I. (2006) *Global Biopiracy: Patents, Plants, and Indigenous Knowledge*. Vancouver: UBC Press.

Montoya, M. (2007) Bioethnic conscription: Genes, race and Mexicana/o ethnicity in diabetes research. *Cultural Anthropology*, 22(1), pp. 94–128.

———. (2011) *Making the Mexican Diabetic: Race, Science, and the Genetics of Inequality*. Berkeley and Los Angeles: University of California Press.

Moreno de Caro, C. (2004) Proyecto de Ley 2 de 2004 por el cual se reglamenta la producción y comercialización de organismo modificados genéticamente con destino al consumo directo tanto de humano como de animales. *Gazeta del Congreso* 409. Bogotá D.C.

Muelas Hurtado, L. (1998) Acceso a los recursos de la biodiversidad y pueblos indígenas. In: Flórez, A. M. (ed.), *Diversidad biológica y cultural: Retos y propuestas desde América Latina*. Santa Fé de Bogotá D.C.: Grupo Ad Hoc Diversidad Biológica / Instituto Latinoamericano de Servicios Legales Alternativos (ILSA)—Grupo Semillas / Instituto de Gestión Ambiental (IGEA) / Proyecto Implementación Convenio sobre Diversidad Biológica (WWF), pp. 171–180.

Need, A. C. and Goldstein, D. B. (2009) Next generation disparities in human genomics: Concerns and remedies. *Trends in Genomics*, 25(11), pp. 489–494.

Neel, J. V. et al. (1980) Genetic studies on the Ticuna, an enigmatic tribe of Central Amazonas. *Annals of Human Genetics*, 44(1), pp. 37–54.

Ng'weno, B. (2007a) *Turf Wars: Territory and Citizenship in the Contemporary State*. Stanford: Stanford University Press.

———. (2007b) Can ethnicity replace race? Afro-Colombians, indigeneity and the Colombian multicultural state. *Journal of Latin American and Caribbean Anthropology*, 12(2), pp. 414–440.

Ordóñez Vásquez, A.; Durán Franch, C.; and Bernal Villegas, J. E. (2006) *Bibliografía anotada: 25 años de investigación en el Instituto de Genética Humana, Facultad de Medicina, Pontificia Universidad Javeriana*. Bogotá D.C.: Pontificia Universidad Javeriana.

Ospina-Duque, J. et al. (2000) An association study of bipolar mood disorder (type I) with the 5-HTTLPR serotonin transporter polymorphism in a population isolate from Colombia. *Neuroscience Letters*, 292(3), pp. 199–202.

Páez Pérez, C. and Freudenthal, K. (1943–1944) Grupos sanguíneos de los indios sibundoy, santiagueños, kuaiker e indios y mestizos de los alrededores de Pasto. *Revista del Instituto Etnológico Nacional*, 1(2), pp. 411–415.

Pardo Rojas, M. (2004) Hitos de la investigación social, histórica y territorial en el Pacífico afrocolombiano. In: Pardo Rojas, M.; Mosquera, C.; and Ramírez, M. C. (eds.), *Panorámica afrocolombiana: Estudios sociales en el Pacífico*. Bogotá D.C.: Instituto Colombiano de Antropología e Historia, pp. 11–25.

Pardo Rojas, M. and Escobar, A. (2004) Movimientos sociales y biodiversidad en el Pacífico colombiano. In: Santos, B. de S. and García Villegas, M. (eds.), *Emancipación social y violencia en Colombia*. Bogotá D.C.: Grupo Editorial Norma, pp. 283–322.

Pineda Camacho, R. (1984) La reivindicación del indio en el pensamiento social colombiano. In: Arocha Rodríguez, J. and Friedemann, N. (eds.), *Un siglo de investigación social: Antropología en Colombia*. Bogotá D.C.: Etno, pp. 197–251.

———. (1995) Colombia étnica. In: *Tierra profanada: Grandes proyectos en territorios indígenas de Colombia*. Santa Fé de Bogotá D.C.: Disloque Editores, pp. 1–37.

———. (1997) La constitución de 1991 y la perspectiva del multiculturalismo en Colombia. *Alteridades*, 7(14), pp. 107–129.

———. (2001) Colombia y el reto de la construcción de la multiculturalidad en un escenario de conflicto. In: Cepeda Espinosa, M. J. and Fleisner, T. (eds.), *Multiethnic nations in developing countries / La pluralidad étnica en los países en vía de Desarrollo.* Friburgo: Institut du Fédéralism Fribourg Suisse, pp. 1–74.

Pinedo Vidal, M. (2001) Proyecto de Ley 151 de 2001 por medio del cual se modifican los Códigos Civil y Penal en lo referente a la aplicación de los métodos de procreación humana asistida, manipulación genética, se dicatan normas sobre el genoma humano de nuestra diversidad étnica, y otras disposiciones. *Gazeta del Congreso* 558. Bogotá D.C.

RAFI (1993) *Patents, Indigenous Peoples, and Human Genetic Diversity* (RAFI Communiqué, May 30). Ottawa and Pittsboro: Rural Advancement Foundation International.

———. (1994a) *The Patenting of Human Genetic Material* (RAFI Communiqué, January 30). Ottawa and Pittsboro: Rural Advancement Foundation International.

———. (1994b) *Gene Boutiques Stake Claim to Human Genome* (RAFI Communiqué, May 30). Ottawa and Pittsboro: Rural Advancement Foundation International.

———. (1995) *Gene Hunters in Search of "Disease Genes" Collect Human DNA from Remote Island Populations.* (RAFI Communiqué, May 30). Ottawa and Pittsboro: Rural Advancement Foundation International.

———. (1996) *New Questions about Management and Exchange of Human Tissues at NIH Indigenous Person's Cells Patented.* (RAFI Communiqué, March 30). Ottawa and Pittsboro: Rural Advancement Foundation International.

———. (1997) *Colombian Indigenous People Negotiate to Get Human Tissue Samples Back.* (RAFI Communiqué, March 28). Ottawa and Pittsboro: Rural Advancement Foundation International.

Ramos, A. R. (2006) The commodification of the Indian. In: Posey, D. A. and Balick, M. J. (eds.), *Human Impacts on Amazonia: The Role of Traditional Ecological Knowledge in Conservation and Development.* New York: Columbia University Press, pp. 248–272.

Rapp, R. (2006) The thick social matrix for bioethics: Anthropological approaches. In: Rehmann-Sutter, C.; Düwell, M.; and Mieth, D. (eds.), *Bioethics in Cultural Contexts: Reflections on Methods and Finitude.* Dordrecht and London: Springer, pp. 314–351.

Reardon, J. (2005) *Race to the Finish: Identity and Governance in An Age of Genomics.* Princeton and Oxford: Princeton University Press.

———. (2007) Democratic mis-haps: The problem of democratization in a time of biopolitics. *BioSocieties*, 2(2), pp. 239–256.

Reichel-Dolmatoff, G. (1959) Urgent tasks of research in Colombia. *Bulletin of the International Committee on Urgent Anthropological and Ethnological Research (IUAES/ UNESCO)*, 2, pp. 50–61.

———. (1967a) A brief field report on urgent ethnological research in the Vaupés Area, Colombia, South America. *Bulletin of the International Committee on Urgent Anthropological and Ethnological Research (IUAES/ UNESCO)*, 9, pp. 53–62.

———. (1967b) Enquêtes ehtnographiques a entreprendre d'urgence (rio Vaupés, Colombie). *Journal de la Société des Américanistes*, 56(2), pp. 320–332.

Reichel-Dolmatoff, G. and Dussan de Reichel, A. (1943–1944) Grupos sanguíneos entre los indios pijao del Tolima. *Revista del Instituto Etnológico Nacional*, 1(2), pp. 507–520.

———. (1966) Proyectos de investigación etnológica en Colombia dentro de un orden tentativo de prioridades por razón de regiones y grupos tribales por estudiar. En: *Documentos del Primer Congreso de Territorio Nacionales*. Documento No. 8. Bogotá D.C.: Congreso de la República.

República de Colombia (1991) *Constitución Política de Colombia*. Bogotá D.C.: Congreso de la República.

———. (1993) Ley 70 (Agosto 27). Bogotá D.C.: Congreso de la República.

———. (2000) Capítulo octavo: Sobre manipulación genética. In: *Ley 599 del 24 de julio 2000* (Código penal). Bogotá D.C.: Congreso de la República.

Restrepo Uribe, E. (2002) Políticas de la alteridad: Etnización de "comunidad negra" en el Pacífico sur colombiano. *Journal of Latin American Anthropology*, 7(2), pp. 2–33.

Rivet, P. (1943) *Los orígenes del hombre americano* (Traducción de José de Recasens). México D.F.: Editorial Cultura.

Salzano, F. M. (2000) James V. Neel and Latin America or how scientific collaboration should be conducted. *Genetic and Molecular Biology*, 23(3), pp. 557–561.

Salzano, F. M. and Bortolini, M. C. (2002) *The Evolution and Genetics of Latin American Populations*. Cambridge and New York: Cambridge University Press.

Salzano, F. M. and Hurtado, A. M. (eds.) (2004) *Lost Paradises and the Ethics of Research and Publication*. Oxford and New York: Oxford University Press.

Sans, M. (2000) Admixture studies in Latin America: From the 20th to the 21st Century. *Human Biology*, 72(1), pp. 155–177.

Santos, R. V. (2002) Indigenous peoples, postcolonial contexts and genomic research in the late 20th Century: A view from Amazonia (1960–2000). *Critique of Anthropology*, 22(1), pp. 81–104.

———. (2006) Indigenous peoples, bioanthropological research, and ethics in Brazil: Issues in participation and consent. In: Ellison, G. T. H. and Goodman, A. H. (eds.), *The Nature of Difference: Science, Society and Human Biology*. London: Taylor & Francis Books, pp. 181–202.

Santos, R. V. and Maio, M. C. (2004) Race, genomics, identities and politics in contemporary Brazil. *Critique of Anthropology*, 24(4), pp. 347–378.

———. (2005) Anthropology, race, and the dilemmas of identity in the age of genomics. *História, Ciências, Saúde-Manguinhos*, 12(2), pp. 1–22.

Shiva, V. (1997) *Biopiracy: The Plunder of Nature and Knowledge*. Boston: South End Press.

Silva-Zolezzi, Irma et al. (2009) Analysis of genomic diversity in Mexican Mestizo populations to develop genomic medicine in Mexico. *Proceedings of the National Academy of Science of the United States of America*, 106(21), pp. 8611–8616.

Skloot, R. (2010) *The Immortal Life of Henrietta Lacks*. New York: Crown Publishers.

Stoler, A. L. (1995) *Race and Education of Desire: Foucault's History of Sexuality and the Colonial Order of Things*. Durham and London: Duke University Press.

———. (2002) *Carnal Knowledge and Imperial Power: Race and the Intimate in Colonial Rule*. Berkeley and Los Angeles: University of California Press.

Sunder Rajan, K. (2006) *Biocapital: The Constitution of Postgenomic Life*. Durham and London: Duke University Press.

TallBear, K. (2007) Narratives of race and indigeneity in the Genographic Project. *The Journal of Law, Medicine & Ethics*, 35(3), pp. 412–424.

Taylor, I. (Director) (1995) *Gene Hunters*. London: Television Trust for the Environment / ZEF Productions / Channel Four (Great Britain) / Films for the Humanities (54 minutes).

Tierney, P. (2000) *Darkness in El Dorado: How Scientists and Journalists Devastated the Amazon*. New York: W.W. Norton & Company.

Ulloa, A. (2005) *The Ecological Native: Indigenous Peoples' Movements and Eco-Governmentality in Colombia*. New York and London: Routledge.

von Heine-Geldern, R. (1959) The International Committee on Urgent Anthropological and Ethnological Research. *American Anthropologist*, 61(6), pp. 1076–1078.

Wade, P. (1993) *Blackness and Race Mixture: The Dynamics of Racial Identity in Colombia*. Baltimore: Johns Hopkins University Press.

———. (1997) *Race and Ethnicity in Latin America*. London: Pluto Press.

Waldby, C. and Mitchell, R. (eds.) (2006) *Tissue Economies: Blood, Organs, and Cell Lines in Late Capitalism*. Durham and London: Duke University Press.

Wang, S. et al. (2007) Genetic variation and population structure in Native Americans. *PLoS Genetics*, 3(11), pp. 2049–2067.

Wang, S. et al. (2008) Geographic patterns of genome admixture in Latin American mestizos. *PLoS Genetics*, 4(3), pp. 1–9.

Wilkie, T. (1995) *Gene Hunters* (Booklet). London: Television Trust for the Environment / ZEF Productions / Channel Four (Great Britain) / Films for the Humanities.

Yunis Turbay, E. (2006) *¿Por qué somos así? ¿Qué pasó con el mestizaje? Análisis del mestizaje*. Bogotá D.C.: Editorial Bruna.

3

The Biological Nonexistence versus the Social Existence of Human Races: Can Science Instruct the Social Ethos?

Telma S. Birchal and Sérgio D. J. Pena

Introduction

In the past, the belief that human "races" had substantial and clearly delimited biological differences contributed to justify discrimination, exploitation, and atrocities. In fact, a clear concept of race has never been clearly established by biology or ethnology, being rather an assumption in theories like that of Gobineau, who claimed the initial diversity of humanity as an a priori datum. The notion of "race"[1] was imported from the common sense to science—and if such notion was not confirmed by the facts, neither was it denied by them.

Recently, however, the advances of the molecular genetics and the sequencing of the human genome have allowed a detailed examination of the correlation between the human genomic variation, biogeographical ancestry, and the physical appearance of people, and these showed that the labels previously used to distinguish races do not have biological importance. It may seem easy to distinguish phenotypically a European from an African or an Asian, but such ease disappears completely when we look for evidence of these racial differences in genomes. In spite of that, the concept of race persists, qua social and cultural construction, as a way of favoring cultures, languages, beliefs, and emphasizing the differences between groups with different economic interests.

In this chapter, we will approach some aspects of the conflict between the social and the biological views on race and analyze it connected to the philosophical question of the relation between science and ethics. We will begin by bringing the scientific evidence that supports the thesis that, from the biological point of view, human races do not exist (AAA, 1998). Next, we will look into the peculiar situation of Brazilians, in whom the broad admixture of genes between three different founding continental groups—Amerindians, Europeans, and Africans—produced a weak correlation of color (a race correlate) with ancestrality. Consequently, in Brazil, color, socially perceived, has little or no biological consequence.

After that, we will broadly discuss the relation between science and ethics, focusing on the "is-ought problem," first articulated by David Hume, or, in other words, into the question of the relation between facts and values. We will reflect upon the specific problem of the incorporation of the teachings of genetics in the ethos of society, in two directions.

First, we intend to show that the analysis of the case in hand actually reinforces Hume's claim—we cannot make claims about what *ought to be* simply on the basis of what *is*—but recognizes, at the same time, the need to establish a relation between science and ethics. An analysis of the phenomenon of racism shows that although science cannot establish the foundations for values, scientific knowledge can bring elements to clarify and inform ethical decisions. Thus, one of its most important tasks is to avoid errors and prejudice, promoting a liberating role in the exercise of moral choices. Being aware that the causes of intolerance "are much deeper than mere ignorance or prejudice" (Lévi-Strauss, 1971), we understand that the contribution of science to fight racism is limited but also is fundamental in a society that values scientific knowledge.

Second, we will argue for the idea that the scientific fact of the non-existence of races must be absorbed by society and incorporated into its convictions and moral attitudes, in the sense of reinforcing the idea of equality and the opposition to the different forms of hierarchy between people or human groups. Using the Brazilian context, we will bring some considerations to the polemics: since races do not exist from a biological point of view, would it lead to a moral consequence that the social use of the concept of race should be banned?

We will conclude suggesting that a desirable and coherent posture would be the valuing of each individual's singularity instead of his identification as a member of racial or "color" groups.

The Biological Nonexistence of Human Races: Scientific Facts

In this section, we review human genomic variability using a historical perspective from the recent origin of modern humans in Africa to their spread to colonize the whole planet. It is shown that the worldwide distribution of human diversity reflects such evolutionary history. In other words, the genetic relatedness of human populations can be better predicted by geography than by ethnic labels. This suggests that ethnic labels will not prove to be an adequate replacement for the appropriate genotyping of patients.

Moreover, we propose that rather than thinking about populations, ethnicities, or races, we should focus on the unique genome of the particular individual, which is structured as a mosaic of polymorphic haplotypes with diverse genealogical histories (Paabo, 2003). This shifts the emphasis from populations to persons. We should strive to see each individual as having a singular genome and a unique life history, rather than try to impose on him/her characteristics of a group or population.

The Origin and Dispersion of Anatomically Modern Humans

Anatomically modern *Homo sapiens sapiens* is a very young species on our planet. Several lines of evidence suggest its single and recent origin, 150,000–195,000 years ago, in Africa. The first is the observation of a genetic diversity in Africa larger than in any other continent. The interpretation of this finding is that a more ancient population, such as Africa, would have more time to accumulate genetic variability. Besides, there was a considerable founder effect in the "out-of-Africa" migration ca. 60,000 years ago. Genetic trees furnish the second line of evidence. Beginning with the seminal work of R. Cann et al. (1987), essentially all studies based on human mitochondrial DNA (mtDNA) have produced a tree in which the first bifurcation separates African populations from those of other continents. Likewise, trees built from autosomal markers, X-chromosome markers and Y-chromosome markers present similar topology. A third compelling line of evidence for a recent African origin of modern humankind is the observation that geographic distance—not genetic distance—from East Africa along likely colonization routes is highly correlated with the genetic diversity of human populations (Prugnolle et al., 2005). Finally, we have dating based on the molecular clock (i.e., the known regularity of neutral mutation along time) that shows a coalescence time for mtDNA lineages around 150,000–200,000 years ago.

Until recently, we were missing critical fossil evidence that could back up the "out-of-Africa" hypothesis for the origin of humankind. In 2003,

T. D. White et al. (2003) described fossilized hominid crania found in Herto, Ethiopia, that have been isotopically dated to 160,000–150,000 years before present (YBP). These hominids, who have been named *Homo sapiens idaltu* (*idaltu* means old in afar, the language of Ethiopia), are morphologically intermediates between ancient hominid fossils and fossils with modern morphology, and thus they were proposed as candidates for being the immediate ancestor of *Homo sapiens sapiens*. Also recently, two skulls found in the Kibish in southern Ethiopia and bearing phenotypic characteristics of anatomically modern humans (AMH) have been dated as having 195,000 years (McDougall et al., 2005). The anatomy and the antiquity of these fossils provide powerful evidence that humankind emerged recently in Africa.

Sometime, probably within the last 70,000 years, AMH left Africa and colonized other continents, decimating and replacing in their trajectory Neanderthals (*Homo sapiens neandertalensis*) and other archaic populations of *Homo sapiens*. According to this scenario, all human beings living presently on earth share a recent African ancestor.

The Races of Humanity: Typological Paradigms

A simple morphological inspection of people from different regions of earth will reveal an apparent paradox: we are at the same time very similar and yet very different. Indeed, there are great similarities among humans: the corporal plan, the erect posture, the thin skin, and the relative scarcity of body hair distinguish us from the other primates. On the other hand, there are significant morphological variations among individuals: height, skin pigmentation, hair texture, facial features, and so on. In special, each one of us has a morphological individuality: our relatives and our friends can identify us in a crowd without any hesitation. This morphological variety can be described at two different levels. The first is at the interpersonal level, the diversity that distinguishes a person from other within a population and that is intimately connected with personal identity. The second is at the interpopulational level, that is, the morphological diversity that characterizes different human groups, especially in different continents.

The latter diversity is very relevant, because historically it has served as a basis for the typological division of humankind into "races." The most influent proposition in this sense was that of the German anthropologist Johann Friedrich Blumenbach (1752–1840). In the 1795 edition of his book *De generis humani varietate nativa* ("On the natural varieties of humanity") he divided all humans into five groups, defined both by geography and appearance: Caucasians (the light-skinned people of

Europe, Middle East, Central Asia, North Africa, and India), Mongolians (East Asia), Ethiopians (the dark-skinned people of Africa), Americans (Amerindians), and Malays (Oceania). The name Caucasian has a double origin: first because in the opinion of Blumenbach the perfect human type was found in the Mountains of Georgia, in Central Asia, and second because he believed that that region had been the cradle of humankind (Gould, 1994). Blumenbach's classification persists to our day, is spite of the fact that we now know that it is impossible to separate humanity in biologically significant categories, independent of the criterion adopted.

In Blumenbach's classification of humankind into different races and in subsequent attempts to do so from Ernst Haeckel in 1868 to Carleton Coon in 1962, the major emphasis was placed on the "interracial" diversity and considerable less importance was given to "intraracial" variability. In a recent conference at the University of California, Richard Lewontin (2004) made the relevant observation that a mark of prejudice and racism is exactly this vision of humanity only in interpopulational terms, that is, the inability to recognize in other "racial" groups the individuality of each person. This is often verbalized as: "They seem all equal to me, but we are all different from each other." When you deny the individuality of members of other groups, you dehumanize and objectify them.

The description of the interpersonal and interpopulational morphological variabilities belongs to the sphere of appearances, the phenotypic world. Subjacent to the observable morphological individuality there is indeed an absolute genomic individuality. However, contrary to the typological paradigm, the genomic representation of the variability between the human groups of different continents—the so-called human "races"—is very small. The physical characteristics that distinguish continental groups apparently represent morphological adaptations to the physical environment, thus being the products of natural selection acting on a very small number of genes. Let us now examine the evidence for these statements, starting with the latter.

Geography and Phenotypic Appearance

J. H. Relethford (1994) showed that only 11–14 percent of human craniometrical diversity occurs between different continents, while 86–89 percent occurs between individuals within regions. When the same author partitioned the variability in skin pigmentation, he observed a very different picture: 88 percent of variation occurred between geographical regions, and only 12 percent within regions (Relethford, 2002). This discrepancy can be explained because skin pigmentation appears to be a special phenotypic feature subject to natural selection. Indeed, two opposing selective factors have been proposed to influence the adaptation of skin pigmentation to prevailing levels of environmental UV radiation: lack of synthesis of vitamin D3 when UV radiation is insufficient and destruction of folate when it is excessive (Jablonski and Chaplin, 2000, 2002). There is an excellent correlation between levels of UV radiation and levels of skin pigmentation worldwide.

The degree of skin pigmentation is determined by the amount and the type of melanin in the skin, and these in turn are apparently determined by a small number of genes (four–six) of which the melanotropic hormone receptor appears to be the most important (Sturm et al. 1998; Rees, 2003). This is an insignificantly small number of genes amid the 20,000–25,000 structural genes in the human genome.

Likewise, external phenotypic features such as nose format, lip thickness, and hair color and texture most likely represent adaptations to environmental conditions and/or are influenced by sexual selection. Just like the pigmentation of skin, these phenotypical features depend on few genes. In summary, these iconic "race" features correlate well with the continent of origin, but depend on variation in an insignificantly small portion of the human genome. We may say that in this sense, race is skin deep. Yet, human societies have constructed elaborate systems of privilege and oppression based on these insignificant genetic differences (Bamshad and Olson, 2003).

Genetic Markers

Subjacent to the abundant human morphological individuality there are abundant levels of metabolic, molecular, and genomic variability (Pena et al., 1995). With the explosion of knowledge derived from the "DNA revolution," our understanding of human genomic diversity has increased exponentially in the past few years (Cavalli-Sforza, 1998).

If for the moment we ignore migrations, the dynamic of variation in allele frequencies of genomic markers is governed by the interactive forces of mutation, selection, and genetic drift. Although nongeneticists have clear concepts of mutation and selection, the phenomenon of genetic drift is lesser known and deserves special elaboration.

The name genetic drift is given to the purely random variation in allele frequencies along time occurring as a sampling effect. The set of alleles of a given generation is not an exact copy of the preceding generation, but is a random sample of it, and as such is subject to statistical fluctuations, like a genetic lottery. Like in every random sample, there is a variance that is inversely proportional to the size of the sample. When the effective size of a population is small, especially when there are drastic populational reductions (bottlenecks) or when a small group leaves the original population and colonizes a different region (founder effect), we can observe important allele frequency variations from one generation to the next (Tischkoff and Verelli, 2003). Occasionally alleles can be fixed (reach frequency one) or removed from the population (reach frequency zero) purely as a consequence of stochastic effects (Kimura, 1989). Because the vast majority of DNA markers used in the study of human diversity is selectively neutral, genetic drift along with mutation pressure is of paramount importance in shaping the distribution of human diversity.

It is relevant to note that while the mutation rate is specific for each locus, thus varying in different parts of the genome, genetic drift depends on the demography and evolutionary history of populations, thus affecting equally all neutral loci in the genome (Luikart et al., 2003).

Partition of Human Genetic Variability

In 1972, Richard Lewontin tested scientifically the notion of the existence of human races as typological entities by partitioning human genetic variability into three additive components: the variability between continents (i.e., between "races"), the variability between population groups within continents, and the variability between individuals within populations. To accomplish that, he researched in the available literature the allele frequencies of 17 classical genetic polymorphisms. He then grouped the populations into eight "racial" continental groups: Africans, Amerindians, Australian aborigines, East Asians (Mongoloids), South Asians, Indians, Oceanians, and Caucasians. The results came as a surprise: 85.4 percent of the allelic diversity occurred within-population groups, 8.3 percent among populations of the same "race," and only 6.3 percent among the so-called races. These data could be better understood using a thought experiment: imagine that a nuclear cataclysm destroys all people on earth with the exception of Africans. In that case, 93 percent of human genetic diversity would be preserved. If only one African population remained, for instance the Zulus from South Africa, we would still maintain about 85 percent of human genetic variability!

This work was criticized because it made use of some polymorphisms of selective value, such as the Duffy blood group that is related to resistance to malaria. Thus, there could be established a correlation between certain allele constellations. However, the pattern is maintained even when neutral DNA polymorphisms are used. For instance, very recently we (Bastos-Rodrigues et al., 2006) undertook a study of worldwide variability using the Human Genome Diversity Project-Centre D′ Etude du Polimorphisme Humain (HGDP-CEPH) Diversity Panel (1,064 individuals from 52 populations) with a set of 40 biallelic short insertion-deletion polymorphisms (indels). These are slow-evolving markers not subject to natural selection, and thus the distribution of their variability reflects only the forces of migration and genetic drift. The 52 different populations originated from 7 geographical regions: Europe, Middle East, Central Asia, East Asia, Oceania, the Americas, and sub-Saharan Africa. With the 40-indel battery, we observed that 85.7 percent of the allelic diversity occurred within population groups, 2.3 percent among populations of the same "race," and only 12.1 percent among the so-called races. These numbers are very similar to those observed by Lewontin with classical markers. Other authors have also obtained similar figures using biallelic markers (Barbujani et al., 1997; Watkins et al., 2003).

These studies illustrate what can be called the "population paradigm" of human genome diversity. E. Mayr (1982) defined population thinking and contrasted it with the previous typological, essentialist thinking that we mentioned above. According to him,

> Population thinkers stress the uniqueness of everything in the organic world. What is important for them is the individuals, not the type. They emphasize that every individual is uniquely different form all others. There is no "typical" individual and mean values are abstractions.... The differences between biological individuals are real, while the comparison of groups of individuals are man-made inferences.

Geographical Correlations of Human Diversity

N. A. Rosenberg et al. (2002) famously typed the same HGDP-CEPH Diversity Panel (1,064 individuals from 52 populations) that we described above with 377 autosomal microsatellites. Later they enlarged this set to 993 markers with no major change in conclusions (Rosenberg et al., 2005). In the sample they observed a total of 4,199 alleles, 47 percent of which were present in all world regions studied—only 7 percent of the alleles were observed in a single region, which in almost every case was Africa. These results indicate that most of human genetic diversity is shared among all regions of the world and is absolutely compatible with the recent single origin of modern humankind in Africa.

When Rosenberg et al. partitioned the variability, they observed that 93–95 percent was contained within populations, a figure considerably higher than the one observed by Lewontin (1972) or us (Bastos-Rodrigues et al., 2006). Indeed, L. Excoffier and G. Hamilton (2003) observed that the level of within-population variance observed by Rosenberg et al. (2002) was larger than other worldwide studies and attributed this to the fact that the authors had not used a stepwise mutation model, the most appropriate for microsatellite studies. Not taking homoplasy into account can depress the among-regions variance component (Barbujani et al. 1997; Flint et al., 1999). If one associates the relatively high mutation rate of microsatellites (Leopoldino and Pena, 2003) with the possibility of size constraints for their growth, different populations would tend to approach a common allelic distribution for these markers (Romualdi et al., 2004).

In the same study, Rosenberg et al. (2002) decided to ascertain the capacity of these selectively neutral microsatellites to distinguish structure in human genetic diversity without assigning them a priori to any population or geographical region. For such they used a computer program called *Structure,* which uses a Bayesian method, that tries to estimate for each individual in the sample the proportion of his/her genome that originates in a given cluster, which in turn is estimated from allele frequencies (Pritchard et al., 2000). The estimation procedure is done with different and growing number K of clusters (K = 2, 3, 4, etc.), which has to be input. For each value of K the program produces a posterior probability.

Rosenberg et al. (2002) showed that the maximum posterior probability occurred at K = 5. Then, the clusters produced by the program corresponded to five great regions, namely: (1) sub-Saharan Africa, (2) East Asia, (3) America, (4) Oceania, and (5) a cluster encompassing Europe, North Africa, Middle East, and Central Asia. The study did not show any advantage in invoking a sixth cluster. It is relevant to know that under the same study protocol and using the same samples we obtained virtually identical results with our set of 40 short indels (Bastos-Rodrigues et al., 2006).

There is an apparent and superficial correspondence between the results of this study and the five human races defined in the eighteeth century by Blumenbach, that is, Ethiopian, Mongoloid, American, Malay, and Caucasian respectively. Indeed, some researchers and the press—including a recent book by the well-known science journalist Nicholas Wade (2006)—have claimed that Rosenberg's study helped reestablish the notion of human races on modern scientific grounds. However, such views are erroneous and cannot withstand close scrutiny.

First of all, we should note that even though the analysis using the *Structure* program was undertaken without a priori population classification, the sampling strategy was clearly population based. Moreover, the sampled groups were few and distant from each other (Kittles and Weiss,

2003). D. Serre and S. Paabo claimed to show using simulation studies that if sampling had been done with individuals on a geographical grid rather than being population based, the clustering effect would be much diminished (Serre and Paabo, 2004).

Second, we should realize that if we choose any two individuals in the same cluster they will only be on an average 4–5 percent more similar than if any of them were compared to an individual from any other cluster. This may lead to statistically significant clustering but that does not mean that it is biologically significant. In other words, individuals from the same geographical region and individuals from different geographical regions are almost equally different!

Third, every racial classification has been based on the wrong typological idea that races were very different from each other and very internally homogeneous. That is not the picture that emerges from Rosenberg's data, which on the contrary shows very heterogeneous clusters barely different from each other. Indeed, Rosenberg's data and similar studies can be used as a strong argument that human races do not exist (Templeton, 1999).

We have already mentioned the work of F. Prugnolle et al. (2005) who showed that geographic distance—not genetic distance—from East Africa along likely colonization routes is highly correlated with the genetic diversity of human populations. The same authors (Manica et al., 2005) later used the data of Rosenberg et al. (2002) to show that pair-wise geographical distances across landmasses constitute a far better predictor of neutral allele sharing than ethnicity! In other words, the distribution of neutral human diversity reflects human evolutionary history.

The observation that allele sharing between human populations worldwide decays smoothly with increasing physical distance is most compatible with a model of colonization of the world based on serial founder effects (Ramachandran et al., 2005; Liu et al., 2006). Fitting of the data to such a model translates into an estimate of the initial expansion of modern humans from East Africa ca. 56,063 ± 5,678 years ago, from an ancestral effective population source of around 1,000 individuals. Thus, it appears that modern *Homo sapiens* remained in Africa for a long time after his origin 160,000–195,000 years ago.

Populations and Individuals

As we saw above, population thinking stresses the uniqueness of individuals within populations. However, as pointed out by R. Caspari (2003) we should realize that such theoretical population thinking may be quite distinct from population studies in practice. In fact, many contemporary anthropologists and geneticists conceptually deal with populations in the same manner as the previous generations dealt with races (Caspari, 2003). Contrary to Mayr's population paradigm, what is important for them is the population and not the individual! Thus, they divide humanity into populations, which can be defined on the basis of geography, culture, religion, physical appearance, or whatever other criterion that is convenient. It appears that such division of humanity into populations does not constitute the most appropriate approach to deal with human variation. Treating people, for instance, of the European population and African population, as separate categories for genetic studies tends to contribute to the public perception that the primary difference between these ways of defining populations is biological (Foster and Sharp, 2004). This view confounds several issues and obscures the important fact that Europeans are genealogically related to Africans, having evolved as an offshoot of the latter.

The human evolutionary history is remarkably short and the worldwide geographical distribution of genetic traits is basically due to dispersal, with ensuing mutation, selection, and genetic drift. In essence, the genetic diversity observable in Europe, Asia, Oceania, and the Americas is merely a subset of the variation found in Africa (Yu et al., 2002). As pointed out by Paabo (2003), from a genomic perspective we are all Africans, either living in Africa or in quite recent exile outside of Africa.

Thus, rather than conceptualizing humans as belonging to defined populations, it might be more appropriate to think of them as 6 billion individuals who have different degrees of pair-wise relatedness along genealogical lines. This has the advantage of firmly connecting our vision of human genetic diversity to our evolutionary history and, as we will see shortly, to genome structure.

This view becomes particularly clear when seen from the perspective of what we call lineage markers. These are the uniparental maternal (mtDNA) and paternal (nonrecombinant regions of the Y chromosome—NRY) polymorphisms, which are haploid and do not recombine. As such, blocks of genes (haplotypes) are transmitted to the next generations and remain unaltered in the matrilineages and patrilineages until a mutation supervenes. The mutations that have occurred and reached high frequencies after the dispersion of modern man from Africa can be specific to certain regions of the globe and can serve as geographical markers. The mtDNA and the NRY provide complementary information that can trace back to several generations in the past. Two observations are of extreme importance here. The first is that if two individuals have the same mitochondrial or Y chromosome haplotype, they are genealogically related along that line, independent of which population they formally belong to.

The second is that the genealogical matrilineage to which an individual belongs to is completely independent of his patrilineages. For instance, our studies have shown that while white Brazilians carry almost exclusively Y-chromosomal lineages phylogeographically related to Europe, two thirds of them have mtDNA lineages phylogeographically related to Amerindians or Africans (Alves-Silva et al., 2000; Pena et al., 2000; Carvalho-Silva et al., 2001). In other words, most Brazilians have these two genomic compartments of different phylogeographical origin and thus are genealogical mosaics.

What about the diploid biparental nuclear genome? The same kinds of genealogical principles that apply to lineage markers also apply in theory to nuclear genes, whose multigeneration transmission routes involve both genders (Avise, 2000). In terms of formal theory, the major difference from uniparental markers is a fourfold adjustment required to account for the larger effective population size of autosomal alleles. This leads to corresponding fourfold longer coalescent times. A second, less important, difference is that in autosomes, besides mutations, lineages can change because of intragenic (or intrahaplotype, see below) recombination events.

In the past few years, it has become evident that much of the human genome is composed of haplotypic blocks ("hapblocks") where polymorphic markers (especially single nucleotide polymorphisms—SNPs) are strongly associated over distances as large as 170 kb (Paabo, 2003; Tischkoff and Verelli, 2003; Wall and Pritchard, 2003). The discussion of the origin of these haplotype blocks is beyond the objective of this review. Suffice to say that probably the length of haplotype blocks is influenced by both demographic factors (which is certainly responsible for most of the variation of block sizes among populations) and genomic factors, especially the existence of recombination hot spots (Zhang et al., 2003; Greenwood et al., 2004). The existence of such hapblocks has high significance for the feasibility of mapping disease genes by marker association studies, since each block can be defined by typing only four–five SNPs. Thus, the number of SNPs needed to achieve fine genomic screen might be reduced from millions to a few hundred thousand (Tischkoff and Verelli, 2003).

We can then envisage the human genome as composed of hundreds of thousands of small genomic blocks of high linkage disequilibrium (like the mtDNA or Y chromosome), each one with its own pattern of variation and genealogical origin. Rather than thinking about populations, ethnicities, or races, we may then consider the genome of any particular individual as a mosaic of variable haplotypes (Paabo, 2003). This is the Variable Mosaic Genome (VMG) paradigm, which completely shifts the focus from populations to individuals. In other words, the paradigm emphasizes human individuality rather than membership in populations. We should strive to see each person as having an individual genome, rather than try to impose on him/her characteristics of the group or population. This is ideally suited to the practice of medicine, since in the office, doctors evaluate and treat individual patients and not populations or races.

Conclusions

We reviewed three models of human genetic structure. The first, preponderant on the nineteenth century and the first half of the twentieth century envisaged humanity as partitioned into well-defined races. This typological model erroneously visualized races as being very different from each other, but internally homogeneous. The consequences of this model were racism, prejudice, and discrimination, leading to the Nazi movement and apartheid.

Beginning in 1930–1940, the typological paradigm was replaced by a model that focused on populations, which were viewed as internally heterogeneous and differed only in allele frequencies. Although theoretically correct, in practice, populations were confused with races, and this second model has been associated with continuing racism and prejudice and recently led to the unfortunate development of the strategy of "race-targeted drugs." This strategy involves polygenic constellations that differ little, perhaps at most twofold or threefold among fairly ill-defined populations (such as African Americans). These polygenes influence (and not determine) the pharmacogenetics and pharmacodynamics of some drugs, and there is ample room left for epistasis or modification by individual genotypes. Thus, such policy is equivalent to playing a game of probabilistic black boxes and erroneously calling it personalized medicine.

Three recent scientific developments have triggered a shift to a much-needed new paradigm: (1) the demonstration of absolute genome individuality in humans; (2) the genetic and paleontological demonstration of a recent and unique origin for modern man in Africa; and (3) the discovery that the human genome is structured in haplotype blocks. The new paradigm is genealogical in nature and based on human evolutionary history—it stresses individuality rather than membership in populations. According to it, we can envisage the human genome as composed of hundreds of thousands of small genomic blocks of high-linkage disequilibrium, each one with its own pattern of variation and genealogical origin. Under this model, ideas, such as that of human races or "race-targeted drugs" become meaningless.

Such paradigm resonates well with many strands of thought in social science. For instance, recently the Nobel Laureate Amartya Sen wrote a book entitled *Identity and Violence* (2006) emphasizing the necessity of humans to define their identity multidimensionally and not according to any single major overarching criteria, such as color, race, or creed that would inevitably lead to divisiveness and conflict. The new genealogical paradigm that emphasizes individual uniqueness is the only one that does not constrain the plural definition of personal identity. Moreover, such paradigm is in perfect alignment with the concept that human rights apply to the individual and not to groups. As famously expressed by US Supreme Court Justice Anthony Kennedy (1995): "At the heart of the Constitution's guarantee of equal protection lies the simple command that government must treat citizens as individuals, not as components of a racial, religious, sexual or national class."

Color and Ancestry in Brazilians

In Brazil, not withstanding relatively large levels of genetic admixture and a myth of "racial democracy," there exists a widespread social prejudice that seems to be particularly connected to the physical appearance of the individual (Harris and Kotak, 1963). Color (in Portuguese, *cor*) denotes the Brazilian equivalent of the English term race (*raça*) and is based on a complex phenotypic evaluation that takes into account, besides skin pigmentation, also hair type, nose shape, and lip shape (Telles, 2002). The reason why the word "color" is preferred to "race" in Brazil is probably because it captures the continuous aspects of phenotypes (Telles, 2002). In contrast with the situation in the United States, there appears to be no racial descent rule operational in Brazil, and it is possible for two siblings differing in color to belong to completely diverse racial categories (Harris and Kotak, 1963).

Based on the criteria of self-classification of the 2000 census of the Instituto Brasileiro de Geografia e Estatística (IBGE), the Brazilian population was then composed of 53.4 percent whites, 6.1 percent blacks, and 38.9 percent brown ("pardos" in Portuguese). How do these numbers correlate with genomic ancestry?

Inferences about the European and African Genomic Ancestral Roots

Using a panel of genetic polymorphisms that display large differences in allelic frequencies (> 0.40; these polymorphisms are called ancestry-informative markers, or AIMs for short) between Europeans and Africans,

E. J. Parra et al. (1998) showed that, at a population level, it was possible to estimate with great precision the degree of European and African ancestry among North Americans. We decided to ascertain whether this same panel of markers would be capable to estimating, on an individual level, the degree of African ancestry in Brazilians. For that, we selected ten of the best AIMs used in the American study.

With the purpose of verifying the individual discrimination power of this set of ten AIMs we initially genotyped small samples of individuals from the northern part of Portugal and from the island of São Tomé, located in the Gulf of Guinea, on the west coast of Africa (Parra et al., 2003). These population sources were chosen because they are geographically related to the European- and African-population groups that participated in the peopling of Brazil. A complete individual discrimination between the European and African genomes was obtained. It was thus clear that the ten-allele set of Parra et al. (1998) was highly efficient and provided reliable individual discrimination between European and African genomes.

Our initial Brazilian sample was composed of 173 individuals from a southeastern rural community, clinically classified according to their color (white, black, or intermediate) with a multivariate evaluation based on skin pigmentation in the medial part of the arm, hair color and texture, and the shape of the nose and lips. When we compared the African genomic ancestry values assessed for these individuals, we observed that the groups had much wider ranges than those of Europeans and Africans, and that there was very significant overlap between them. This indicated that in Brazil there is significant dissociation of color and genomic ancestry, that is, at an individual level it was not possible to infer the ancestry of a Brazilian from his/her color (Parra et al., 2003).

To corroborate these findings, we undertook a second investigation based on data from 12 forensic microsatellites that had been utilized to estimate the personal genomic origin for each of the 752 individuals from the city of São Paulo, belonging to different Brazilian color categories (275 whites, 192 browns, and 285 blacks) (Pimenta et al., 2006). The genotypes permitted the calculation of a personal likelihood-ratio estimator of European or African ancestry. Again we observed great overlaps among color categories of Brazilians. This was confirmed quantitatively using a Bayesian analysis of population structure that did not demonstrate significant genetic differentiation between the color groups. These results corroborate and validate our previous conclusions using AIMs.

If we consider some peculiarities of Brazilian history and social structure, we can construct a model to explain why color should indeed be a poor predictor of African ancestry (Parra et al., 2003). Nowadays most

Africans have black skin, genetically determined by a very small number of genes that were evolutionarily selected in adaptation to the tropical and subtropical climate. Thus, if we have a social race–identification system based primarily on phenotype, such as occurs in Brazil, we can classify individuals on the basis of the presence of certain alleles at a small number of genes that have impact on the physical appearance, while ignoring all of the rest of the genome. Assortative mating based on color, which has been shown by demographic studies to occur in Brazil, will produce strong associations among the individual components of color. Indeed, we detected the presence of such positive associations at highly significant levels in a southeastern Brazilian population (Parra et al., 2003). On the other hand, we expect that any initial admixture association between color and the AIMs will inevitably decay over time because of genetic admixture. It is easy to see how this combination of social forces could produce a population with distinct color groups and yet with similar levels of African ancestry.

Genomic Studies of the Amerindian, European, and African Ancestral Roots

The two studies mentioned above did not take into account the Amerindian contribution to the Brazilian population. To achieve that we needed new polymorphic markers that would be sensitive to all three ancestries.

We screened the database of 2,000 human diallelic short indels characterized by J. L. Weber et al. (2002) and identified 40 polymorphisms that fulfilled the following criteria: widespread chromosomal location in the human genome, increasing amplicon sizes that allowed multiplex polymerase chain reaction (PCR) amplification and electrophoretic analysis and allele frequency close to 0.5 in the European population. We used these 40-indel markers to study worldwide human genome variation, namely all the samples in the CEPH-HGDP Diversity Panel (Cann et al., 2002), composed of 52 populations originated from 5 geographical regions: America, sub-Saharan Africa, East Asia, Oceania, and a cluster composed of Europe, Middle East, and Central Asia (Cann et al., 2002). We obtained a distance matrix of the 52 populations using the Reynolds genetic measure. On visualization, using Multidimensional Scaling analysis, we obtained a very adequate graphical representation that showed five widely separated clusters corresponding to Africa, Oceania, East Asia, America, and a central Europe-Middle East-Central Asia group (Bastos-Rodrigues et al. 2006).

Next, we submitted the genotypes of the Amerindians, Europeans, and sub-Saharan Africans of the CEPH panel (the three Brazilian ancestral roots) to *Structure* version 2.1 (Pritchard et al., 2000), a Bayesian software that uses multilocal genotypes to infer the structure of population and group individuals on the basis of their genotypes even without any prior information on the population origin of each sampled individual. The program produced a triangular plot in which the three different populations clustered in different vertices with no overlap. European individuals had on an average 94.6 percent European ancestry, sub-Saharan Africans had on an average 96.5 percent sub-Saharan African ancestry, and Amerindians had on an average 94.8 percent Amerindian ancestry (Pena et al., 2009).

Inferences about the Amerindian, European, and African Genomic Ancestral Roots Among Brazilians

As shown in the previous section, Amerindian, African, and European samples of CEPH can be used to define a triangular landscape with ancestry-specific vertices on which we can plot the results of *Structure* analyses of Brazilian samples. We obtained plots for Brazilian self-declared white individuals from several regions of Brazil, as well as from Brazilian self-declared black individuals from the city of São Paulo (Pena et al., 2009). Most of the white individuals of all the regions examined clustered in the "European" vertex of the triangle plot, although a proportion of them were scattered throughout the triangle area. All average European contributions for white individuals were above 0.700, with a maximum of 0.819 in Southern Brazil (a region of heavy European immigration), and a minimum of 0.709 in Minas Gerais. When we compared the regions pairwise using a Monte Carlo resampling strategy, however, we could not find statistical significance (Pena et al., 2009). Black individuals from the city of São Paulo showed very high individual variation in their biogeographical ancestry, as indicated by a large spread of data points in the triangular graph (Pena et al., 2009). On the average they had an average degree of African ancestry slightly below 50 percent, which was significantly different from white individuals of the same region and also from individuals from sub-Saharan Africa.

An interesting observation is that the extent of Amerindian ancestry is relatively low (range 0.092–0.147), and not statistically different among white individuals from different geographical regions and also between white and black individuals from São Paulo.

Conclusions

Many authors used historical, sociological, and anthropological method-ology to analyze the origins of Brazilians: Paulo Prado in *Retrato do Brasil* (published in 1927), Sérgio Buarque de Holanda in *Raízes do Brasil* (published in 1933), Gilberto Freyre in *Casa Grande & Senzala* (published in 1933), and Darcy Ribeiro in *O Povo Brasileiro*. We have used new molecular genetics tools for the same purpose.

The data presented in this review demonstrate that currently available DNA test can provide an important molecular confirmation of the proposals of the authors mentioned above and also are capable of providing new valuable insights into the process of genetic formation and structure of the Brazilian people.

Studies with uniparental markers in both white and black Brazilians demonstrate strong directional mating between European males and Amerindian and African females, which agrees with the known history of the peopling of Brazil since 1500. These data reveal that the genomes of most Brazilians are mosaic, having mtDNA and NRY with different phylogeographical origins.

Studies with autosomal biparental markers reveal very elevated levels of genetic admixture between the three ancestral roots. However, it is also evident that there was an important population effect of the program of "whitening" of Brazil promoted through the immigration of ca. 6 million Europeans in the roughly one hundred year period after 1872. This manifests itself both in a predominant (>70 percent) European genomic ancestry in Brazilian whites regardless of geographical region and on a high average European genomic ancestry (37.1 percent) in Brazilian blacks.

The correlation between color and genomic ancestry is very imperfect: at an individual level one cannot safely predict the skin color of a person from her level of European, African, and Amerindian ancestry nor the opposite. Regardless of their skin color, the overwhelming majority of Brazilians have a high degree of European ancestry. Also, regardless of their skin color, the overwhelming majority of Brazilians have a significant degree of African ancestry. Finally, most Brazilians have a significant and very uniform degree of Amerindian ancestry!

The high ancestral variability observed in whites and blacks suggest that each Brazilian has a singular and quite individual proportion of European, African, and Amerindian ancestry in their mosaic genomes. Thus, the only possible basis to deal with genetic variation in Brazilians is not as members of color groups, but on a person-by-person basis, as 190 million human beings, with singular genome and life histories.

Science and Ethics

We shall, then, consider the question: what kind of practical or moral consequences could be authorized from scientific discoveries concerning the notion of race? Should we keep race as a social category even if it has no genetic meaning?[2] These questions take us directly to the philosophical problem of the relation between science and ethics.

The Fact-Value Problem I

In order to think about the terms of this relation, we shall take a view of science that, even if it is debatable, can work as a starting point: sciences intend to be the knowledge of *what is*, they deal with the research and presentation of reality. In this point, one must, undoubtedly, take into account well-known criticisms, like that raised by Hilary Putnam about the positivist's ideal: science is not, as believed, a mere presentation of facts, because the notion of experience in science covers hypotheses and aprioristic constructions; nor it is axiologically neutral, for it is moved by values, like simplicity, coherence, and objectivity (Putnam, 2004)—in fact objectivity, if it does not describe science's effective reality, is its supreme value. Once that is said, though, it still makes sense to claim that there is a distinction between the scientific and the moral point of views because sciences turn themselves to the positive reality, while ethics has a different interest.

Also, in a first approach, the field of ethics or morality can be defined as the realm of what *ought to be* and what is good. We can identify as a general characteristic of the field of ethics its *prescriptive or normative* character. The several historical ethics consist of commandments, customs, and laws that display how things *ought to be* based in some general conception of the good. We can, then, in broad lines, set the limits between science and ethics, relating the former to the world of *facts* and to what *is*, and the latter to the world of *values* and to what *ought to be*. Thus, the language of ethics is essentially *prescriptive*, whereas the language of science is essentially *descriptive*; ethics is *evaluative*, concerning the good and bad, science is not.

Leaving apart the difficult subject of the relation between morals and truth (can a moral judgment be true or false?),[3] it is a dramatic fact of our experience that the moral domain is an object of conflict and very unlikely to bend toward universal agreement. Centuries ago, Plato (who strongly believed in moral truths) had already distinguished between the two domains, when he wrote that if there is dispute and hatred between the gods, this is not due to the disagreements about, for example, the weight or measure of something or other matters about which the overcoming of doubt is possible, since one weighs or measures the object. However, about the just and the unjust, the beautiful and the ugly, and good and evil—themes so linked to the wishes and interests that an agreement is made impossible—gods end up in war (Plato, Euthyphro, 7a–9a). The moral domain is linked to our desires in a way that science is not.

If we accept that these differences make sense, we will agree with the classic formulation of this problem by the philosopher David Hume who, in the eighteenth century, pointed the distinction between facts and values, when he stressed the logical error of deducing an "ought" statement form an "is" statement.[4] Hume forbids us to pass from the "is" (fact) to the "ought" (value) *without giving some reason for it, that is, ought* statements cannot be immediately deduced from *is* statements *because they express different relations.* In other words, there's a "gap" between facts and values, so that facts do not have the power to generate norms, and all transition from fact to value must be justified.

We can understand the so-called Hume's law as a criticism of naturalism as a conception in ethics that, since the ancients, has prescribed to "follow nature" as its maxim. One cannot say that something is good just *because* it's natural. One will always need first to justify *why* the natural is good. And although most of the philosophers do justify their assertions, the reasons they present are not merely descriptive, but appeal to moral or metaphysical criteria (what is not a problem per se, but reinforces the idea that we cannot move from mere facts to values). An example of that is the Stoic philosophy itself: the principle "follow nature" only works because it is based on the assumption that nature is a good order and arranged by the Providence. More recently, Spencer's Social Darwinism claimed that society should "follow nature," where nature stands for the realm of competition leading to the "survival of the fittest." The assumption is that competition brings progress, which is good. In both cases, the idea of nature is arbitrarily invested of a value that justifies the principle (Stewart-Williams, 2004).

More simply, it seems obvious that we cannot directly go from "facts" to "values" or from nature to ethics because, if we did it, we could, for example, justify violence because it is "in our genes." In a pure Humean spirit, the contemporary French philosopher Comte-Sponville writes: "Truth never rules, nor says what must be done or prohibited. Truth does not obey…and that's why it is free. Nor it commands, and that's why we are free. It is true that we will die: this does not condemn life, nor justifies murder" (Comte-Sponville, 2001).

The first claim—"truth never rules"—means that no description of things is immediately prescriptive (Hare, 1972), for human action is not restricted to simply repeat things as they are in the world, and this is, finally, the most essential sense of freedom. The second claim—"truth does not obey"—recognizes the existence of a domain that, having objectivity as its highest directive, does not surrender to human being's wishes and dreams—we will return to this point—only to itself. This last claim, however, is much more related to an ideal than to an assertion that science is neutral: Obviously, sciences are linguistic and cultural products, and, as such, are subject to injunctions and values of their context. Nevertheless, science *ought to* search for objectivity, without subjecting itself to any interest, religion, or political beliefs.

Facts and Values II: Science's Contribution to Ethics

So far we have presented ethics as a domain that articulates prescriptions and values, and we have left the world of facts for science. However, it is also true that ethics relates itself with the world of facts, and in several ways.

The most evident is that historically existent norms, laws, habits, and values belong to the world of facts; the questions about right and wrong come from a living experience that is caught, among others, by the scientific perspective. Hence one can talk, for example, about the "Protestant ethics," and even treat it as an object of an empirical science, as it was done by Max Weber's sociology. In the same way as it does with the physical objects, science takes morality as an empirical reality: neuroscience locates, in the brain, the sites of moral capacities, and evolutionist psychologists describe the moral instincts that constrain our moral judgments (Hauser, 2006). However, as it happens to scientific knowledge in general, this knowledge of the various aspects of moral experience is not immediately prescriptive. As for the Protestant ethics, we should still judge if it is good or bad, something that Weber, in his role of a scientist, did not intend to do; as for the discovery of the biological conditioning of behaviors, it does not tell us anything definitive about the attitude one must have in relation to them, and most of the moral psychologists agree about that (see, for example, Greene, 2003).

Nevertheless, scientific knowledge, in the form of presentation of facts, has profound ethical implications and is important in clarifying and informing ethical choices. This informing role of science has at least two dimensions: the first, which we could call "prudential," refers to the capacity of scientific knowledge to show the consequences of our choices, so that the decisions are taken reasonably, and the discussions are carried out advisedly.[5] The second, which we could call "critical," has to do with the ability of science to show the falsity of ideas, beliefs, and theories; this way, it can carry out its liberating task, for it has the power to put errors and prejudice away.

We have already had the opportunity to say that truth only obeys itself, that is, it does not coincide either with our desires or with our dreams. Now we claim that just because of that, knowledge becomes liberating. In other words: although facts or reality are morally neutral in themselves, we are not indifferent to them. It was not with indifference that humanity learned that the Earth turns around the Sun, or that human beings are the fruit of a natural selection. Science is, in its vocation, indifferent to the human desires—even to those of the powerful—it simply doesn't obey, and this has an ethical implication. That's why, we believe, we *ought to take* scientific data into consideration in ethical issues. The famous three wounds in human narcissism of which Freud tells us—Copernicus's heliocentric theory, Darwin's theory of evolution, and psychoanalysis itself—had profound consequences for ethics, and we would say in a humanizing and liberating direction. More recently, for example, the discovery of chemical factors associated to the working of mind and emotions, allowed a different understanding of mental disturbances and deviating behaviors.

This critical dimension of science is also pointed out by K. W. Appiah as a capacity to take a distance from ourselves.[6] This distance—we believe—is very important to make us able to see ourselves better.

In short, science provides new elements that have to be considered in evaluating our values. It cannot, however, be taken for the origin or the source of moral commandments, or in other words, it's not immediately prescriptive. The sciences bring elements that contribute to ethical reflection and enlarge the field on which one can exercise one's freedom, but that are not able to generate ethics by themselves. Last but not least, science is itself considered one of the greatest values of the Western society.

Conclusion: Ethnocentrism, Racism, and the Individual View of Man

We shall, then, go back to the notion of race and ask what kind of moral consequences could be authorized from scientific discoveries we have described.

The notion of race finds its origin in the secular experience of the confrontation with the other, whose difference appears sensitively, evidently, say, superficially. The meeting with the different has historically generated different reactions, and among these reactions several forms of ethnocentrism. According to anthropologists, ethnocentrism is widespread among human groups—it is "the best distributed thing in the world" (Henaf, 2000, 348)—and according to some biologists, it is part of our natural evolved dispositions (Hauser, 2006, 211–214). Although we cannot do it in the limits of this chapter, it would be important to distinguish between ethnocentrism and racism;[7] we will just remark that ethnocentric disposition can generate racism, and racism, in turn, develops the notion of race, necessary for its reinforcement.

Anyhow, one is not justified in identifying some so-called biological dispositions—if they do exist—to a deterministic account of a fixed human nature that would necessarily lead humanity to discrimination or racism (Pena, 2008, 56–57). And even if we accept that we have some evolved dispositions to ethnocentrism, it's good to remember that something is not moral just for being natural or for having strong cultural roots. Along with ethnocentrism it's also possible to find, in different moments of the history of humanity, several examples of criticism of this kind of conviction or practice. Usually, a nonethnocentric point of view demands a kind of consciousness that moves away from common sense and from immediate experience. Shall we remember Michel de Montaigne, who in the sixteenth century wrote about the natives of Brazil, claiming that to understand them we should get away from the "vulgar opinion, and that we are to judge by the eye of reason, and not from common report" (Essays, Book I, chapter 31).

Science can contribute to the construction of this perspective. Today, genetics asserts that, from the biological, strictly scientific point of view, human races do not exist. In other words, the apparent and immediate displayed—color and features difference—is not the essential and true, demanding the reinterpretation of common experience, in the same way that, in the sixteenth century, astronomy did with the apparent movement of the sun. In the domain of facts, science establishes that the biological notion of race is a mistake. This knowledge, morally neutral in itself, has some ethical consequences that we could express in Popper's terms: science can say *what is not*, although it cannot say what *ought to be*. In the case in question, it clearly forbids the appeal for a biological foundation of racism, and makes clear that they were grounded on pseudoscientific beliefs. That is, it provides new arguments to supply a nonracist posture. Hence the possibility to denounce that not only racism, but the concept of race itself is charged with ideology and has its origins in relations of power and domination (Munanga, 2004). In other words, it is possible to establish that racism is not predicated on biological existence of races, but on their social existence. That is, races exist because they are in people's head, but are not in people's head because they exist (Kaufman, 1999).

But still a nonracist posture is an ethical conviction and not a scientific knowledge. To make it clear, let us consider the following situation:

One can believe in the biological existence of races and still be nonracist. Hypothetically, we could think that, if biology had discovered deep differences (objective facts) between human beings, even though, it wouldn't follow that these differences must be interpreted as signs of inferiority or superiority (value judgments). A distinction between racialism[8] and racism can be established. *Differences* (as factual descriptions) can be stated keeping the horizon of *equality* (a normative ideal). We can go back again to Montaigne, who does not fail in recognizing a lot of differences between the people from the Old World and the New World, but does not understand the latter as inferior. Or to Wilberforce and his group, who fought for the abolishment of the traffic of slaves in England (1831), and probably believed in important biological differences between Europeans and Africans. More recently, in his two articles for the UNESCO, Claude Lévi-Strauss shows that, even if we could talk about races, we would have to appeal to disputable values to interpret the biological difference in terms of cultural inferiority or superiority. This means that the facts brought by science are not a necessary condition for standing for the value of equality.

So we meet again the idea that science cannot be the source of the values or moral commandments. Racism should be fought not because it's not scientific, but because it is a discriminating and oppressing attitude. Nevertheless, scientific data can reinforce or not reinforce values that have already been given previously. To use an image from Peirce, a nonracist posture (scope of values) does not follow deductively from the nonexistence of races (scope of facts), like chain links follow one another, but scientific discoveries contribute with strong thread to weave the rope of antiracism, together with so many other threads, from anthropology, from the equality and fraternity ideals, and so on. Or, on the contrary: the racist speech is weakened without the support of scientific data. Science cannot say what we ought to do, but it does forbid us to walk on certain paths: when it says "what is not," it delivers us from errors and prejudices. That is clearly what is happening with the recent genetic discoveries.

It is, then, up to society, informed by science, to recognize a value in the fact of biological nonexistence of races, that is, invest it with a meaning beyond itself. The genetic information about human beings helps to build a more universal image of the human kind.

*   *   *

Now we arrive at our final point. It's far from consensual that, since there's no sense in talking about races from the biological point of view, we should also avoid the social use of this category. Many people wouldn't agree that all use of the concept of race is equal to racism or discrimination; the concept of race could be taken in a positive sense, for example, for reinforcing group identities or as the basis for social "affirmative action." They will argue that, if there's a social existence of races, this has to be considered in social and political action. And it's important to say that those who stand for a "color conscious" position or for affirmative actions are guided by the very same final values as those who stand for a "color blind" position: equality.

Does the social concept of race, stressing group particularities, contribute to the construction of a more equal and fair society as claims A. Mosley (2005) or does it "retribalize people into ethnic or other fixed-identity groups," introducing new kinds of exclusivism, as claims J. B. Elshtain (1998)? There's no a priori answer to this question, and here social and historical sciences can help, bringing other kind of facts to inform our choices.

For the specific case of Brazil, we claim that the discoveries of genetics must be taken into account in the discussion about affirmative action. From the genetic perspective, we have seen that, independently from their color, the vast majority of Brazilians have simultaneously an expressive degree of African, European, and Amerindian ancestry. The genome of each Brazilian is a highly individual and variable mosaic, formed by contributions from the three ancestral roots (Suarez-Kurtz and Pena, 2006). Thus, it does not make sense to talk about Afro-descendants or Euro-descendants. Furthermore, because of the poor correlation between color and ancestry, it does not make sense to talk about "populations" of white Brazilians or black Brazilians. Thus, the only way of dealing scientifically with the genetic variability of Brazilians is individually, as singular and unique human beings in their mosaic genomes and in their life histories. From the medical point of view, this awareness led us to propose that the concept of race must be abolished from Brazilian medicine (Pena, 2005).

For deciding about the social use of the category of race in the quota policy, we should also look for contributions from social sciences. Thomas Sowell (2004) has shown, in his book *Affirmative Action around the World*, that the world experience with quotas has increased the level of racialization of society. We strongly believe we should avoid this effect in Brazilian society. Biology contributes effectively to a nonracialist conception of mankind. And in Brazil, the consciousness of the weak correlation between color and ancestry meets the utopian wish of a nonracialist society, "blind to colors," where the individual singularity is valued and celebrated. And this should be searched ethically and socially, as Martin Luther King thought when he said, in his famous speech "I have a dream," on August 28, 1963: "*I have a dream that my four children will one day live in a nation where they will not be judged by the color of their skin but by the content of their character.*"

In his book *Against Race,* Paul Gilroy (2000) brilliantly demonstrated that this is feasible. Then, we must concentrate our efforts in this direction. When implementing well-intentioned programs of affirmative action, to push the necessary social changes forward, the government must be careful not to stir up artificial and arbitrary tensions and divisions among the people of Brazil, a country where, essentially, we are all equally different.

Notes

First published in *Revista USP*, 68, pp. 10–21, 2006.

1. In theory, the term "race" should come always into quotation marks, since it refers to a biologically nonexistent entity. However, we will not use them in this text because it becomes tiresome.

2. In an important paper, Philip Kitcher (2007) has posed a similar question (Does "race" have a future?), but his answers are quite different from ours. We are grateful to Steve Clarke for this indication.

3. This is a much disputed question in philosophy. A sound answer is given Karl Popper, pointing the difference between natural and moral laws: "A natural law is describing a strict, unvarying regularity which either in fact holds in nature (in the case the law is a true statement) or does not hold (in this case is false).… A normative law, whether it is now a legal enactment or a moral commandment, can be enforced by men. Also, it is alterable. It may perhaps be described as good or bad, right or wrong, acceptable or unacceptable; but only in a metaphorical sense can it be called 'true' or 'false,' since it does not describe a fact, but lays down directions for our behaviour" (Popper, 1950, 58–59).

4. "In every system of morality, which I have hitherto met with, I have always remark'd, that the author proceeds for some time in the ordinary way of reasoning, and establishes the being of God, or makes observations concerning human affairs; when on a sudden I am surprised to find, that instead of usual copulations of propositions, *is*, and *is not*, I meet with no proposition that is not connected with an *ought*, or an *ought not*. This change is imperceptible ; but is, however, of the last consequence. For as this ought, or ought not, expresses some new relation or affirmation, 'tis necessary that it shou'd be observ'd and explain'd ; and at the same time that a reason should be given, for what seems altogether inconceivable, how this new relation can be a deduction from others, which are entirely different from it" (Hume, 1992, 468).

5. We thank prof. Alcino Bonella for the suggestion that we include this point.

6. "Ethicists sometimes fear that science will alienate us from our values; yet the exercise of estrangement—of testing our values by trying to distance

ourselves to them—is one that humanists have pursued as vigorously as scientists" (Appiah, 2008, 159).

7. It is known by the anthropological studies that it is usual that a certain tribe applies the label "human" only to itself, thus implying that the other tribes do not take part in human nature. Lévi-Strauss (1971) claims that certain dosage of ethnocentrism is essential for the construction of the identity of a people or a group, what is done through the differentiation and even the opposition in relation to others. Recognizing with Lévi-Strauss the need of a certain ethnocentrism to keep the identity of a group, Rorty writes: "We would rather die than to be ethnocentric, but ethnocentrism is precisely the conviction that one would rather die than to share certain beliefs" (Rorty, 1991, 203).

8. We have used the word *racialist* to nominate those who believe in the existence of races, differently from *racist*, which refers to those who make judgments of value and establish hierarchies among "races."

References

AAA (American Anthropological Association). (1998) American Anthropological Association Statement on "race". Available from http://www.aaanet.org /stmts/racepp.htm.

Alves-Silva, J. et al. (2000) The ancestry of Brazilian mtDNA lineages. *American Journal of Human Genetics*, 67, pp. 444–461.

Appiah, K. W. (2008) *Experiments in Ethics*. Cambridge, MA: Harvard University Press.

Avise, J. C. (2000) *Phylogeography: The History and Formation of Species.* Cambridge, MA: Harvard University Press.

Bamshad, M. J. and Olson, S. E. (2003) Does race exist? *Scientific American*, 289, pp. 78–85.

Barbujani, G. et al. (1997) An apportionment of human DNA diversity. *Proceedings of the National Academy of Sciences of the United States of America*, 94, pp. 4516–4519.

Bastos-Rodrigues, L. et al. (2006) The genetic structure of human populations studied through short insertion-deletion polymorphisms. *Annals of Human Genetics*, 70, pp. 658–665.

Cann, H. M. et al. (2002) A human genome diversity cell line panel. *Science,* 296, pp. 261–262.

Cann, R. et al. (1987) Mitochondrial DNA and human evolution. *Nature*, 325, pp. 31–36.

Carvalho-Silva, D. R. et al. (2001) The phylogeography of Brazilian Y-chromosome lineages. *American Journal of Human Genetics*, 68, pp. 281–286.

Caspari, R. (2003) From types to populations: A century of race, physical anthropology, and the American Anthropological Association. *American Anthropologist*, 105, pp. 65–76.

Cavalli-Sforza, L. L. (1998) The DNA revolution in population genetics. *Trends in Genetics*, 14, pp. 60–65.

Comte-Sponville, A. (2001) *a Small Treatise on the Great Virtues*. New York: Metropolitan Books.

Elshtain, J. B. (1998) Democracy and the politics of difference. In: Etizioni, A. (ed.) *The Essencial Communitarian Reader*. Oxford: Rowman & Littlefield Publishers, pp. 259–268.

Excoffier, L. and Hamilton, G. (2003) Comment on "Genetic structure of human populations." *Science*, 300, p. 1877.

Flint, J. et al. (1999) Minisatellite mutational processes reduce Fst estimates. *Human Genetics*, 6, pp. 567–576.

Foster, M. W. and Sharp, R. R. (2004) Beyond race: Towards a whole-genome perspective on human populations and genetic variation. *Nature Review Genetics*, 5, pp. 790–796.

Gilroy, P. (2000) *Against Race: Imagining Political Culture beyond the Color Line*. Cambridge, MA: Harvard University Press.

Gould, S. J. (1994) The geometer of race. *Discover*, 15, pp. 65–69.

Greene, J. (2003) From neural "is" to moral "ought": What are the moral implications of neuroscientific moral psychology? *Nature Reviews*, 4, pp. 847–850.

Greenwood, T. A. et al. (2004) Human haplotype block sizes are negatively correlated with recombination rates. *Genome Research*, 14, pp. 1358–1361.

Hare, R. (1972) *The Language of Morals*. Oxford: Clarendon Press.

Harris, M. and Kotak, C. (1963) The structural significance of Brazilian categories. *Sociologia*, 25, pp. 203–208.

Hauser, M. (2006) *Moral Minds: How Nature Designed Our Universal Sense of Right and Wrong*. New York: Ecco.

Henaf, M. (2000) *Claude Lévi-Strauss et l'anthropologie structural*. Paris: Points.

Hume, D. (1992) *a Treatise of Human Nature*. Oxford: Oxford University Press.

Jablonski, N. G. and Chaplin, G. (2000) The evolution of human skin coloration. *Journal of Human Evolution*, 39, pp. 57–106.

———. (2002) Skin deep. *Scientific American*, 287, pp. 74–81.

Kaufman, J. S. (1999) How inconsistencies in racial classification demystify the race construct in public health statistics. *Epidemiology*, 10, pp. 101–103.

Kennedy, A. (1995) *Miller v. Johnson (94-631)—U.S. Supreme Court Decision 515*. Available from http://supreme.justia.com/us/515/900/case.html.

Kimura, M. (1989) The neutral theory of molecular evolution and the world view of the neutralists. *Genome*, 31, pp. 24–31.

Kitcher, P. (2007) Does "race" have a future? *Philosophy and Public Affairs*, 35(4), pp. 293–317.

Kittles, R. A. and Weiss, K. M. (2003) Race, ancestry, and genes: Implications for defining disease risk. *Annual Review Genomics and Human Genetics*, 4, pp. 33–67.

Leopoldino, A. M. and Pena, S. D. J. (2003) The mutational spectrum of human autosomal tetranucleotide microsatellites. *Human Mutation*, 21, pp. 71–79.

Lévi-Strauss, C. (1971) Race and culture. *UNESCO's International Social Science Journal*, 13(4), pp. 608–626.

Lewontin, R. C. (1972) The apportionment of human diversity. *Evolutionary Biology*, 6, pp. 381–398.

———. (2004) The concept of race: The confusion of social and biological reality. Available from: http://www.uctv.tv/library-science.asp?seriesnumber=17.

Liu, H. et al. (2006) A geographically explicit genetic model of worldwide human-settlement history. *American Journal of Human Genetics*, 79, pp. 230–237.

Luikart, G. et al. (2003) The power and promise of population genomics: From genotyping to genome typing. *Nature Review Genetics*, 4, pp. 981–994.

Manica, A. et al. (2005) Geography is a better determinant of human genetic differentiation than ethnicity. *Human Genetics*, 118, pp. 366–371.

Mayr, E. (1982) *The Growth of Biological Thought*. Boston: Belknap.

McDougall, I. et al. (2005) Stratigraphic placement and age of modern humans from Kibish, Ethiopia. *Nature*, 433, pp. 733–736.

Mosley, A. (2005) A defense of affirmative action. In: Cohen, A. and Wellman, C. H. (eds.) *Contemporary Debates in Applied Ethics*. Malden: Blackwell, pp. 43–58.

Munanga, K. (2004) Uma abordagem conceitual das noções de raça, racismo, identidade e etnia. *Cadernos PENESB*, 5, pp. 15–34.

Paabo, S. (2003) The mosaic that is our genome. *Nature*, 421, pp. 409–412.

Parra, E. J. et al. (1998) Estimating African American admixture proportions by use of population-specific alleles. *American Journal of Human Genetics*, 63, pp. 1839–1851.

Parra, F. C. et al. (2003) Color and genomic ancestry in Brazilians. *Proceedings of the National Academy of Sciences of the United States of America*, 100, pp. 177–182.

Pena, S. D. J. (2005) Razões para banir o conceito de raça da medicina brasileira. *História, Ciências, Saúde—Manguinhos*, 12, pp. 321–346.

———. (2008) *Humanidade sem raças?* São Paulo: Publifolha.

Pena, S. D. J. et al. (1995) DNA diagnosis of human genetic individuality. *Journal of Molecular Medicine*, 73, pp. 555–564.

Pena, S. D. J. et al. (2000) Retrato molecular do Brasil. *Ciência Hoje*, 27(159), pp. 16–25.

Pena, S. D. J. et al. (1995) DNA diagnosis of human genetic individuality. *Journal of Molecular Medicine*, 73, pp. 555–564.

Pena, S. D. J. et al. (2009) DNA tests probe the genomic ancestry of Brazilians. *Brazilian Journal of Medical and Biological Research*, 42, pp. 870–876.

Pimenta, J. R. et al. (2006) Color and genomic ancestry in Brazilians: a study with forensic microsatellites. *Human Heredity*, 62, pp. 190–195.

Popper, K. R. (1950) *The Open Society and Its Enemies. I. Plato.* Princeton: Princeton University Press.

Pritchard, J. K. et al. (2000) Inference of population structure using multilocus genotype data. *Genetics*, 155, pp. 945–959.

Prugnolle, F. et al. (2005) Geography predicts neutral genetic diversity of human populations. *Current Biology*, 15, pp. R159–R160.

Putnam, H. (2004) *The Collapse of the Fact/Value Dichotomy and Other Essays*. Cambridge, MA: Harvard University Press.

Ramachandran, S. et al. (2005) Support from the relationship of genetic and geographic distance in human populations for a serial founder effect originating in Africa. *Proceedings of the National Academy of Sciences of the United States of America*, 102, pp. 15942–15947.

Rees, J. L. (2003) Genetics of hair and skin color. *Annual Review of Genetics*, 37, pp. 67–90.

Relethford, J. H. (1994) Craniometric variation among modern human populations. *American Journal of Physical Anthropology*, 95, pp. 53–62.

———. (2002) Apportionment of global human genetic diversity based on craniometrics and skin color. *American Journal of Physical Anthropology*, 118, pp. 393–398.

Romualdi, C. et al. (2002) Patterns of human diversity, within and among continents, inferred from biallelic DNA polymorphisms. *Genome Research*, 12, pp. 602–612.

Rorty, R. (1991) *On Ethnocentrism: a Reply to Clifford Geertz. Objectivity, Relativism and Truth. Philosophical Papers*. Volume I. Cambridge: Cambridge University Press.

Rosenberg, N. A. et al. (2005) Clines, clusters, and the effect of study design on the inference of human population structure. *PLoS Genetics*, 1, p. e70.

Rosenberg, N. A. et al. (2002) Genetic structure of human populations. *Science*, 298, pp. 2381–2385.

Sen, A. (2006) *Identity and Violence: The Illusion of Destiny*. New York: W.W. Norton & Company, p. 215.

Serre, D. and Paabo, S. (2004) Evidence for gradients of human genetic diversity within and among continents. *Genome Research*, 14, pp. 1679–1685.

Sowell, T. (2004) *Affirmative Action around the World: An Empirical Study*. New Haven: Yale University Press.

Stewart-Williams, S. (2004) Darwin meets Socrates. *Philosophy Now*, 45, pp. 26–29.

Sturm, R. A. et al. (1998) Human pigmentation genetics: The difference is only skin deep. *Bioessays*, 20, pp. 712–721.

Suarez-Kurtz, G. and Pena, S. D. J. (2006) Pharmacogenomics in the Americas: Impact of genetic admixture. *Current Drug Targets*, 7, pp. 1649–1658.

Telles, E. (2002) Racial ambiguity among the Brazilian population. *Ethnic and Racial Studies*, 25, pp. 415–441.

Templeton, A. R. (1999) Human races: A genetic and evolutionary perspective. *American Anthropologist*, 100, pp. 632–650.

Tishkoff, S. A. and Verrelli, B. C. (2003) Patterns of human genetic diversity: Implications for human evolutionary history and disease. *Annual Review of Genomics and Human Genetics*, 4, pp. 293–340.

Wade, N. (2006) *Before the Dawn: Recovering the Lost History of Our Ancestors*. New York: Penguin Press.

Wall, J. D. and Pritchard, J. K. (2003) Haplotype blocks and linkage disequilibrium in the human genome. *Nature Review Genetics*, 4, pp. 587–597.

Watkins, W. S. et al. (2003) Genetic variation among world populations: Inferences from 100 *Alu* insertion polymorphisms. *Genome Research*, 13, pp. 1607–1618.

Weber, J. L. et al. (2002) Human diallelic insertion/deletion polymorphisms. *American Journal of Human Genetics*, 71, pp. 854–862.

White, T. D. et al. (2003) Pleistocene *Homo sapiens* from Middle Awash, Ethiopia. *Nature*, 423, pp. 742–747.

Yu, N. et al. (2002) Larger genetic differences within Africans than between Africans and Eurasians. *Genetics*, 161, pp. 269–274.

Zhang, K. et al. (2003) Randomly distributed crossovers may generate block-like patterns of linkage disequilibrium: An act of genetic drift. *Human Genetics*, 113, pp. 51–59.

Part II

Genomics, Genetic Admixture, and Health in South America: Old and New Opportunities and Challenges

4

Admixture Mapping and Genetic Technologies: Perspectives from Latin America

Bernardo Bertoni

Introduction

During the past decades, we have witnessed the progressive growth of information on genetic diseases. The current challenge relates to understanding complex diseases where environment and genetics are interacting within a dense and complex network. The development of new techniques and methodologies are making it possible to dissect and examine the etiology of complex diseases. Every month, from different areas (biochemistry, epidemiology, molecular biology, and genetics) terabytes of data are being generated around the world in relation to these kinds of diseases. But full understanding of the cause and cures of these diseases are still far away, even though we are getting closer. The effort is justified because diseases like cancer, hypertension, diabetes, or even obesity represent a major challenge for the health systems and the economies of the countries. The question examined in this chapter is, how can we understand the idea of admixture mapping as a tool capable of giving new insights in understanding the development of complex diseases? In this chapter, a brief description about, capabilities and limitations of this approach are described, from a specific South American perspective.

Nowadays understanding complex diseases is a hard task. The paradigm a "gene for a specific disease," can only be justified in a few cases. It is increasingly becoming apparent that the phenotype that we call "disease" represents a complex network of relationships between environment, the organism, and its genes. Within this network, gene-gene interactions are one among many of the numerous pathways by which this disease is understood; multiple genes and multiple variants of the genes can be involved in disease development. Some of these genes have minor effects, and others have major effects. These effects can be defined with a detailed phenotype description, which helps to reduce—for analytic purposes—the genetic heterogeneity. But in other cases, individual variants of a gene can produce different phenotypes in different genetic backgrounds as has been found in model organisms (Sinha et al. 2006). Nevertheless, we also need to consider an emerging vision that describes some syndromes as "oligogenic" diseases. The oligogenic diseases are conformed by few genes interacting between each other. An example is the Bardet-Biedl syndrome, a ciliophaty involving retinal degeneration, obesity, hypogonadism, polydactyly, renal dysfunction, and mental retardation. Recently, a study has demonstrated that a significant fraction of the mutations (rare variants) of a gene group have a strong negative mode of action. Meanwhile, a group of common variants in the population interacts with these strong rare variants to modulate disease presentation (Zaghloul et al. 2010). Rare variants are present in extremely low frequencies in the population, which means that few people carry them. On the other hand, common variants have a high frequency, and anyone could be a carrier of the variant. So, the interplay between rare variants and common variants in the population are determining the disease presentation. It is important to keep in mind that the variant frequency in one population (e.g., Europe) is not necessarily the same in other population from other continent. This point is important for the admixture mapping as we will see in next sections.

But the gene architecture is not the only factor that shape complex traits; there are also other non-Mendelian characteristics like extra copies of a DNA segment (Copy Number Variation—CNV). This extra segment can include genes distorting the norm of two gene copy per individual (Mendel laws). However, a genome-wide association study of CNVs gave no conclusive data for the genetic basis of common diseases (Craddock et al. 2010). Nevertheless, chromosome segment repetitions need to be evaluated more carefully in the context of evolutionary selection and their role in metabolic pathways. A strong selection against CNV on key proteins of metabolic pathways was detected in a recent study (Schuster-Bockler, Conrad, and Bateman 2010). But again, selection pressures act in a different way in a population in comparison to another from a different continent. The selection's marks can be detected in present-day populations and can also be seen in recent studies (Akey et al. 2004; Biswas and Akey 2006; Pickrell et al. 2009).

Another aspect of this emerging research paradigm is that we have gene-environment interactions; a field that has experienced a dizzying growth recently aiming to unveil the relationship between alleles and the environment in model organism (Smith and Kruglyak 2008). Besides, connecting the environment and genes we also need to consider the epigenome; DNA modifications without sequence changes can affect the gene expression. Environmental conditions can modify the genome across life, giving individual patterns of DNA modifications, and as a consequence, each individual has a particular pattern of gene expression (Fraga et al. 2005). It is becoming increasingly evident in recent studies that these modifications can be a source of heritable phenotypic variation (Johannes et al. 2009). It is easy to think that some of these "lamarckian" characteristics can be different from one population to another. Recent studies found that African American, Hispanics, and European Americans have a different pattern of genome modifications (Kwabi-Addo et al. 2010; Nielsen et al. 2010), which could be related to ancient adaptations to the environment.

In summary, in considering the complexity at stake in the development of any disease, the genetic level is just one of the multiple levels, which now need to be understood. However, genes are still one of the foundation stones in the architecture of any disease, which is why the community expends much effort in understanding the role of genetic factors.

Many methodologies were developed during the past century to identify the genes thought to be behind a trait or disease. Linkage analysis proved to be a good method for mono-variant diseases/traits. This relied on following the variants shared among affected relatives through generations along with molecular markers. The number of molecular markers typed in these studies is not high. This method emphasizes the effect of rare variants. But, when the increased risk conferred by a variant is small, as in complex diseases, some relatives will be affected because of other causes and will not carry the risk variant. And also, if a variant is common among the population, it can enter the family through multiple founders, erasing clear inheritance patterns. On the other hand, association studies are optimum to detect the common variants. They compare molecular markers in a sample of affected individuals against a sample of normal ones. A significant frequency difference is taken to indicate that the corresponding region of the genome that contains the molecular marker also contains a functional DNA-sequence variant that influences the disease or trait in question (Kruglyak 2008). In the past decades, due to the development of the human genome project, an extremely high number of molecular markers (principally Single Nucleotide Polymorphism—SNP, which usually have two variants or alleles) are available for analysis. Genome-wide association studies (GWAS) emerged as a tool to detect common variants involved in common diseases. They cover the whole genome with a dense map of SNPs (500,000 or more) to look for variant-frequency differences between affected individuals and controls. They require an important number of cases and controls to achieve significant values, but have a higher map resolution and statistical power than linkage analysis. More than 450 papers were published since 2004 (Khoury et al. 2009; Rosenberg et al. 2010), most of them developed in Europe or the United States. Only a small fraction of studies found associations that met the designation of strong evidence to assure association (Lawrence, Evans, and Cardon 2005; Khoury et al. 2009). Many strategies are explored to overcome the problem, but recently, more attention has been given to extend the studies to other populations in an attempt to characterize risk variants. European populations contain only a subset of human variation, which is the result of biological adaptation to the environment and the action of other factors that affected the distribution of risk variants. Therefore, some of these risk variants differ substantially in their relevance to other populations; they are not present or their frequencies differ in the affected individuals. Recently, analyzing an extremely accurate SNP database from European, Asian, and African populations (HapMap) for type 2 diabetes and prostate cancer variants, A. Adeyemo and C. Rotimi (2010) found extremely different risk variants frequencies between populations confirming this long-stated suspicion about the relevancy of different variants.

Despite these limitations, unquestionably GWAS is a powerful method. However, GWAS is not an approach that is viable for many researchers, partly due to cost, especially in Latin America. Even though the costs of the technology behind GWAS are falling, a study that recruits patients and controls represents an investment of millions of dollars. While a few First World grant agencies support programs of this magnitude, there are certainly fewer agencies in Latin America capable of doing this. In past years, admixture mapping has become an alternative method that can be implemented in the region, with less costs but equally effective as GWAS.

Admixture Mapping

R. Chakraborty and K. M. Weiss (1988) introduced the idea of admixture as a new approach to detect risk variants in complex diseases. At first, admixture mapping was hampered due to technical and methodological issues. Later, the explosive growth of a panel of molecular markers (SNPs, microsatellites) and the development of the statistical theoretical frame helped to consolidate this methodology and make it more widely available and usable (Stephens, Briscoe, and O'Brien 1994; McKeigue 1997, 1998; McKeigue et al. 2000; Shriver et al. 2003) .

Admixture mapping is based on the fact that most genetic variations are shared between groups, but the allele frequencies could be different. In the same way, disease-causing variants are known to differ substantially in frequency across populations as was stated before. This is especially relevant for diseases with different incidences across ethnic groups—autoimmune diseases, melanoma, or breast cancer (more common in Europeans) or hypertension and prostate cancer (usually more common in West Africans)—where little is known about the risk variants involved. Admixture mapping is designed to study populations descended from the recent mixing of ethnic groups from multiple parts of the world (urban Latin American populations are a good example). In chromosomal regions containing variants contributing to disease risk, there will be an overrepresentation of ancestry from the parental population that has a higher proportion of risk alleles at the locus (figure 4.1) (Darvasi and Shifman 2005). This method relies on the selection of a group of ancestry informative markers (AIMs) covering the genome. In African Americans or Latin American admixed populations, 2,000 SNPs could be enough to cover the whole genome depending on the characteristic of the admix population. These represent 200 or 300 times less SNPs than in GWAS that translates into a significant reduction in the genotyping cost.

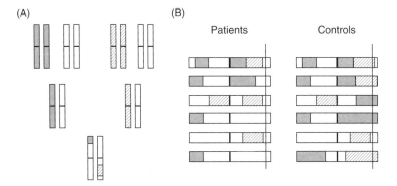

Figure 4.1   Scheme A represents the chromosome mosaic result from a trihybrid admixture process. Each person is represented by a pair of chromosomes. In the first generation, each person belongs to one of three different parental populations (gray, shaded and white respectively). After two generations of mating, each person carries a random combination of parental chromosome segments due to a biological process called recombination. Scheme B shows how an excess of a parental chromosome segment (white segment) is overrepresented in the patient sample.

Geographic Populations

Admixture mapping needs to define the parental populations that contributed to the hybrid population. The assumption of the existence of parental populations underlies the existence of geographical populations as discrete units; however, the topic that generates most discussion is if we can discriminate populations by their genetic load. The discussion is sometimes colored by language that is not precise in biological/genetic terms, making some words or expressions very charged with an excess of "cultural" weight. Here, this discussion focuses on the ability to detect genetic differences between populations; that is, in terms of the number of markers and the samples characteristics necessary to define a population as a genetic discrete unit. When markers and populations are few and the latter come from samples around the world, it is easy to detect differences between them. In this case, it is possible to assign discrete units related to the continents. But if there are few markers and a dense-population sampling scheme used, then it is not possible to detect geographical populations, and there is continuum between continents (Serre and Pääbo 2004). The notion of a continuum between human populations relies on long-term well-stated observation that the fraction of genetic variance corresponding to "between individuals" is greater than "between populations" (Nei and Roychoudhury 1972, 1974). But with the recent advances in genotyping technologies and statistical tools, this approach has been challenged. When the number of markers is increased along with a dense sampling scheme, a continental pattern reappears. An analysis of 377 autosomal microsatellites markers and 1,052 individuals from 52 populations worldwide distributed found significant evidence of a continental clustering (Rosenberg et al. 2002). In a subsequent work, with 993 polymorphisms (783 microsatellites and 210 biallelic markers), the authors confirm that with enough markers with a sufficiently large worldwide sample, individuals can be partitioned into genetic clusters that match geographical continents (Rosenberg et al. 2005). Recently, the results from genetic variability analysis of 197,147 SNPs in European populations have showed that individual genotypic profiles distribute in a close correspondence with geographic distribution: a two-dimension plot of variability resembles the European map (Novembre et al. 2008). However, the possibility of defining discrete units should not be taken as evidence of "biological races." This data is simply saying that two next door neighbors, with families who have been settled for a long time in that place, could share more alleles in common than with others 1,000 km from there. Without a controversial spirit, a PubMed search of the word "race" quoting in the title or abstract for the last five years, has more than

13,400 articles. An important number of articles refer to Hispanics and African Americans, groups of admixed populations defined by cultural/ history parameters. These names are tags that represent heterogenic human groups.

As was stated by Paradies, Montoya, and Fullerton (2007), in association studies it is important not to "racialize" the putative candidate variants or genes found in the population, to avoid possible mistaken generalizations. Probably, in terms of genetic structure, it is better to say Hispanic from California than Hispanic. Or, in our case, Uruguayan is a better tag than Latin American or Hispanic.

In other words, the context of the admixture mapping method, population discrete units work as operational units to define the parental contribution to a particular hybrid population.

Admixed Populations

Admixture is a process common in human populations but perhaps Latin American urban populations are the most studied human hybrid populations. The interests of anthropology and history have fuelled the study of the genetic structure of these populations. Latin America is characterized by a di-trihybrid structure with contribution from European, African, and Native American populations. The process began when Native American, a population isolated for more than 10,000 years, came into contact with European populations. Later, the Africans entered the continent as slaves; a process that lasted until 1850 approximately. The mixing process was not homogeneous and each region had its own characteristics (Bryc et al. 2010). M. Sans (2000) reviews in depth the process in Latin America and the studies done in the past century. Table 4.1 presents the parental contribution in different regions of the continent with data extracted from different resources. The heterogeneity of the contribution depends on the demographic history of each population; an important aspect in the design of admixture mapping studies. Until not so long, Hispanics were considered a homogeneous population in the United States: an assumption that obscured their diverse origins. The term Hispanic comprises Spanish-speaking populations principally Mexicans in the southwest and other non-Mexicans in the east, which gave place to different subpopulations (Hanis et al. 1991; Chakraborty, Fernandez-Esquer, and Chakraborty 1999; Bertoni et al. 2003; Bryc et al. 2010). Table 4.1 also shows differences between regions of Brazil or Argentina. These considerations need to be taken into account, as Latin American populations are not homogeneous even in the same country.

Table 4.1   Admixture estimation from US Hispanics, Mexico, and South American urban populations

Population	European		Native American		African	
	Mean	s.e.	Mean	s.e.	Mean	s.e.
US Hispanic Southwest[†]	0.644	0.016	0.356	0.017		
US Hispanic Southeast[†]	0.933	0.008			0.067	0.008
Mexico City[‡]	0.568	0.061	0.398	0.061	0.034	0.006
Colombia Medellin[*]	0.656	0.031	0.253	0.018	0.091	0.017
Brazil Northeast[‡]	0.719	0.023	0.118	0.016	0.164	0.020
Brazil Rio Grande do Sul	0.696	0.034	0.204	0.099	0.099	0.026
Argentina Catamarca[*]	0.532	0.022	0.437	0.022	0.000	0.006
Uruguay[**]	0.870	0.026	0.103	0.000	0.027	0.010
Argentina Buenos Aires[‡]	0.707	0.026	0.215	0.017	0.078	0.020

Notes: [†]Bertoni et al. 2003, [*]Wang et al. 2008, [**]Hidalgo et al. 2005, [‡] Bertoni et al., unpublished data.

Diseases in Admixed Populations

Admixture mapping relies on the presence of disease prevalence differences between parental populations. When analyzing an individual in a hybrid population that had been originated from more than one ancestral population, the likelihood that this individual can develop a given disease is influenced by the susceptibility to the disease in the ancestral populations. When such an individual carries a hereditary disease, it is more likely that the disease alleles belong to chromosomal segments from the ancestral population with the higher risk. To detect population risk differences, evidence can be found from two sources given below.

(a) The disease risk in human groups from different geographic origins living in the same country or region. Here, the advantage is that the groups share a common environment. So, epidemiological studies in the United States are an important source of data, when exploring a disease in the context of admixture mapping.

(b) The disease risk in the potential ancestral populations (countries, continents) to the hybrid population. Diseases like stroke, lung cancer, prostate cancer, dementia, end-stage renal disease, multiple sclerosis, hypertension, and many more diseases all exhibit a higher morbidity in either Africans or Europeans, when the two ethnically different populations are compared (Smith and O'Brien 2005).

Table 4.2 summarizes data from melanoma and breast cancer taken from international databases (Pisani, Bray, and Parkin 2002; Parkin et al. 2005). For both types of cancer it clearly emerges that South American populations have an incidence rate intermediate between European and African populations. Nonnative American data are available but Eastern Asian population suggests the rates could be low. When compared to Hispanics in the United States, melanoma incidences rates have an intermediate value between European Americans and African Americans (Cress and Holly 1997). With regard to breast cancer, the incidence rate in Hispanics (89.8 per 100,000 inhabitants) is between European Americans (141 per 100,000 inhabitants) and Native Americans (54.8 per 100,000 inhabitants) (Fejerman et al. 2008). In both cases, the admixed population, Hispanic from the United States or urban Latin American populations, have an intermediate rate.

In some cases, it can be difficult to establish these relationships. The risk of diabetes mellitus type 2, varies between population groups in the United States. European Americans have a lower risk than Native American, Latinos, and African American populations (Williams et al. 2000; Martinez-Marignac et al. 2007). By contrast, in Native American populations from Central and South America the diabetes prevalence seems to be lower than in admixed populations, suggesting another factor involved in the development of the disease (Barcelo and Rajpathak 2001). A great limitation in these analyses from different Latin American populations is the partial nature of the data, making relationships more difficult to establish than in North American populations. Moreover, many factors are acting to shape the results. First of all, the measure methods of diabetes are not the same in all the studies and also cultural

Table 4.2 Prevalence (per 100,000 inhabitants) of breast cancer and melanoma in populations from Europe, America, Africa, and Asia

Population	Breast Cancer	Melanoma
Europe	76.5	7.9
North America	99.4	14
South America	46	2.3
Western Africa	27.8	1
Central America	25.9	1.5
Middle Africa	16.5	2.2
Eastern Asia	18.7	0.2

and socioeconomic factors are present, making comparisons a harsh task. However, as was stated before, gene background and gene-environment interaction could also be different in these areas.

The last question relates to the idea of relative recurrence risk in siblings (λ_s), the ratio between the lifetime risk of the disease in the siblings of an affected individual and the lifetime risk of the disease in the general population is a useful way to weigh the familial clustering of the disease (Risch 1990). Even if it is hard to obtain the data necessary for its calculation, λ_s can help to unveil the hereditary component of a disease in admixed populations. It is possible that prevalence can be close to zero in one of the parental population and higher in the other parental population. But, in both populations λ_s could be one, so no family clustering can be detected. However, this fact does not mean that a genetic component is not important in the development of this complex disease. What is happening is that an hypothetical risk variant is absent in one population and extremely common in the other (the variant is common between families and across the generations) (Sawcer et al. 2010). Both situations represent a difficult task. In one case we need to find a family with the disease that share a rare variant, and in the other case we need to plan carefully a considerable amount of samples and molecular markers. This is an important point for the admixture mapping; an admixed population will have an intermediate allele risk frequency. The hereditary information from λ_s will be greater than one, because the risk variant will increase its presence in families with contributions from the ancestral population with the high variant risk frequency.

Admixture Estimation

To test for linkage of a molecular marker to the disease, it is assumed that gamete admixture and locus ancestry can be inferred without uncertainty. Classical methods of admixture estimation are not enough for this task. Statistical methods are now available to infer population stratification and individual admixture using multilocus genotype data. STRUCTURE (Pritchard, Stephens, and Donnelly 2000; Falush, Stephens, and Pritchard 2003) is popular in the literature, but others are available as ADMIXMAP (McKeigue et al. 2000; Hoggart et al. 2003), ANCESTRYMAP (Patterson et al. 2004), and BAPS (Corander, Waldmann, and Sillanpaa 2003). These programs share a Bayesian Markov Chain Monte Carlo (MCM) formulation with two main assumptions: Hardy Weinberg equilibrium and linkage equilibrium between markers within each population. H. Tang et al. (2006) proposes a new method (SABER) and D. Falush et al. (2003)

extended STRUCTURE, both of them take into account the problem of background-linkage disequilibrium.

The programs allow different parameter adjustment that help to adjust the model, based on knowledge of each option and its implications, not always clear for the user. Other methods with a frequentist (maximum likelihood) approach for individual ancestry estimation (Bonilla et al. 2004; Tang et al. 2005) or focusing in principal component analysis (Patterson, Price, and Reich 2006) are available. These methods are less computationally demanding (minutes compared to hours), and if enough markers and the parental populations are known with accuracy the results are not so different with regard to MCM approach (Tsai et al. 2005; Reiner et al. 2007; Aldrich et al. 2008).

Ancestry inference, whether for mapping disease loci or for conducting gene-association studies, is a critical component of genetic analysis in an admixed population.

Mathematical models formalizing admixture mapping can be found in numerous reports (McKeigue 1998; Hoggart et al. 2004; Montana and Pritchard 2004; Zhu, Tang, and Risch 2008). Basically, two tests can be implemented.

(a) Affected only test: The method uses the information of the ancestry estimation at the locus. Near a disease locus, the local mean ancestry of the cases should diverge from the genome-wide mean ancestry of cases. The test measures the divergence of this two means.

(b) Case-control test: In this case, near a disease locus, the local mean ancestry of cases should diverge from the local mean ancestry of controls. The test measures this difference.

In both cases there are different approaches implemented in different software. Parametric approaches will be more powerful if the assumed genetic model is correct, but this is not the case if the genetic model is wrong (Hoggart et al. 2004). On the other hand, nonparametric methods only test for increased ancestry shared among affected individuals (Montana and Pritchard 2004).

Since 2005, an increasing number of published researches are focusing on admixture mapping. X. Zhu et al. (2005) published the results of an African American hypertension study where a *6q24* and *21q21* loci were detected as candidate regions. After this, a continuous effort can be found in bibliography, mainly in African American populations. Table 4.3 summarizes the results from different authors on prostate cancer, cardiovascular disease, or multiple sclerosis. There are many studies

Table 4.3   Chromosome regions or genes described in studies that apply admixture mapping or ancestry association in African American and Hispanic/Latin American populations

Admixture mapping	Chromosome region/ gene /SNPs	Reference
African American		
Hypertension	6q24-6p21.1-21q21	Zhu et al. 2005, Cheng et al. 2010
Prostate cancer	8q24-7q31.31-Bq3B	Freedman et al. 2006, Bock et al. 2009
Cardiovascular disease	IL6SR(rs8192284)	Reich et al. 2007
Multiple sclerosis	1p12-1q21.1	Reich et al. 2005
Peripheral arterial disease	11q13.5	Scherer et al. 2010
Genetic ancestry		
African American		
Lipidic levels		Reiner et al. 2007
Risk of type 2 diabetes		Wassel Fyr et al. 2007
Hispanic/Latin America		
Breast cancer		Fejerman et al. 2008
Obesity		Ziv et al. 2006
Sistemic lupus erythematosus		Seldin et al. 2008
Myocardial disease		Ruiz-Narvaez et al. 2010

detecting genetic ancestry associations with diseases in Hispanic/Latin American and African American populations. Breast cancer, obesity, or diabetes studies indicate that an excess of parental contribution is present in African American or Hispanic or Latin American populations. Even so, in balance, these studies are still in a preparatory stage, collecting information in many cases.

Even though admixed populations can be seen as a potential resource for mapping genes associated to complex diseases, they can also offer other perspectives. That is why hybrid or admixed populations represent a natural experiment for the combination of genomes with different adaptations. Latin American populations, as admixed populations, had their own microevolution history during the past five centuries. Considering them only as the mixture of populations reduces their biology diversity. Hybrid populations allow the observation of known functional genetic variation under a different genetic background. As an example, the presence of fetal hemoglobin (FH) in adult erythrocytes can be explained at least by three loci. These loci—XmnI^G γ site in the β globin gene cluster, the HBS1L-MYB intergenic polymorphism (HMIP) in chromosome 6q, and the *BCL11A* in chromosome 2—account for 50 percent of the variation in the European populations (Menzel et al. 2007; Creary et al.

2009). The expression of FH provides a benefit for patients with sickle cell anemia or beta thalassemia (Perrine 2005). African populations have a higher prevalence of these diseases, therefore African European admixed populations represent a natural experiment of the combination of these two phenotypes. Recently, two admixed populations (African Caribbean and German Africans) were analyzed for HMIP locus. Both show a lower frequency for the allele associated to high levels of FH in the HMIP locus compared to the European population. The most interesting results show that the FH in one African descent population is higher than in the European population (Creary et al. 2009). This finding suggests a different gene interaction responsible for FH in the African descent population than in Europeans.

On the other hand, even if the admixed Latin American populations have few centuries, a recent research found signals of natural selection in the genome of Puerto Ricans (Tang et al. 2007). Two regions, one in chromosome *6p22* and the other in chromosome *11q11* harbor an olfactory gene cluster. A third region in chromosome *8q23.3* seems to be related to a gene (*CSMD3*) with high fetal brain expression (Tang et al. 2007). So, admixture mapping will help not only to uncover susceptibility genes in admixed and parental populations associated to complex diseases, but also to understand the interplay of signaling pathway in metabolic or regulatory routes.

Admixture mapping is offering to Latin American biomedical research an open field of opportunities (Gonzalez Burchard et al. 2005; Salari and Burchard 2007); an affordable methodology with feasible budgets and requirements with the possibility of exploring the implications of findings at clinic, physiology, epidemiology, or even evolutionary level. As a result, although multidisciplinary and/or multinational admixture mapping projects represent a considerable investment of resources for research communities, we can expect a plethora of results in coming years giving new insights in complex diseases.

References

Adeyemo, A. and Rotimi, C. (2010) Genetic variants associated with complex human diseases show wide variation across multiple populations. *Public Health Genomics,* 13, pp. 72–79.

Akey, J. et al. (2004) Population history and natural selection shape patterns of genetic variation in 132 genes. *PLoS Biology,* 2, e286.

Aldrich, M. C. et al. (2008) Comparison of statistical methods for estimating genetic admixture in a lung cancer study of African Americans and Latinos. *American Journal of Epidemiology,* 168, pp. 1035–1046.

Barcelo, A. and Rajpathak, S. (2001) Incidence and prevalence of diabetes mellitus in the Americas. *Revista Panamericana de Salud Publica*, 10, pp. 300–308.

Bertoni, B. et al. (2003) Admixture in Hispanics: Distribution of ancestral population contributions in the continental United States. *Human Biology*, 75, pp. 1–11.

Biswas, S. and Akey, J. M. (2006) Genomic insights into positive selection. *Trends in Genetics*, 22, pp. 437–446.

Bock, C. H. et al. (2009) Results from a prostate cancer admixture mapping study in African-American men. *Human Genetics*, 126, pp. 637–642.

Bonilla, C. et al. (2004) Ancestral proportions and their association with skin pigmentation and bone mineral density in Puerto Rican women from New York City. *Human Genetics*, 115, pp. 57–68.

Bryc, K. et al. (2010) Colloquium paper: Genome-wide patterns of population structure and admixture among Hispanic/Latino populations. *Proceedings of the National Academy of Sciences (U.S.A.)*, 107, Suppl. 2, pp. 8954–8961.

Chakraborty, B. M. et al. (1999) Is being Hispanic a risk factor for non-insulin dependent diabetes mellitus (NIDDM)? *Ethnicity and Disease*, 9, pp. 278–283.

Chakraborty, R. and Weiss, K. M. (1988) Admixture as a tool for finding linked genes and detecting that difference from allelic association between loci. *Proceedings of the National Academy of Sciences (U.S.A.)*, 85, pp. 9119–9123.

Cheng, C. Y. et al. (2010) Admixture mapping of obesity-related traits in African Americans: The Atherosclerosis Risk in Communities (ARIC) Study. *Obesity (Silver Spring)*, 18, pp. 563–572.

Corander, J. et al. (2003) Bayesian analysis of genetic differentiation between populations. *Genetics*, 163, 367–374.

Craddock, N. M. et al. (2010) Genome-wide association study of CNVs in 16,000 cases of eight common diseases and 3,000 shared controls. *Nature*, 464, pp. 713–720.

Creary, L. E. et al. (2009) Genetic variation on chromosome 6 influences F cell levels in healthy individuals of African descent and HbF levels in sickle cell patients. *PLoS One*, 4, e4218.

Cress, R. D. and Holly, E. A. (1997) Incidence of cutaneous melanoma among non-Hispanic whites, Hispanics, Asians, and blacks: An analysis of California cancer registry data, 1988–93. *Cancer Causes Control*, 8, pp. 246–252.

Darvasi, A. and Shifman, S. (2005) The beauty of admixture. *Nature Genetics*, 37, pp. 118–119.

Falush, D. et al. (2003) Inference of population structure using multilocus genotype data: Linked loci and correlated allele frequencies. *Genetics* 164, pp. 1567–1587.

Fejerman, L. et al. (2008) Genetic ancestry and risk of breast cancer among U.S. Latinas. *Cancer Research*, 68, pp. 9723–9728.

Fraga, M. F. et al. (2005) Epigenetic differences arise during the lifetime of monozygotic twins. *Proceedings of the National Academy of Sciences (U.S.A.)*, 102, pp. 10604–10609.

Freedman, M. L. et al. (2006) Admixture mapping identifies 8q24 as a prostate cancer risk locus in African-American men. *Proceedings of the National Academy of Sciences (U.S.A.)*, 103, pp. 14068–14073.

Gonzalez Burchard, E. et al. (2005) Latino populations: A unique opportunity for the study of race, genetics, and social environment in epidemiological research. *American Journal of Public Health*, 95, pp. 2161–2168.

Hanis, C. L. et al. (1991) Origins of U.S. Hispanics: Implications for diabetes. *Diabetes Care,* 14, pp. 618–627.

Hidalgo, P. C. et al. (2005) Genetic admixture estimate in the Uruguayan population based on the loci LDLR, GYPA, HBGG, GC and D7S8. *International Journal of Human Genetics*, 5, pp. 217–222.

Hoggart, C. J. et al. (2003) Control of confounding of genetic associations in stratified populations. *American Journal of Human Genetics*, 72, pp. 1492–1504.

Hoggart, C. J. et al. (2004) Design and analysis of admixture mapping studies. *American Journal of Human Genetics*, 74, pp. 965–978.

Johannes, F. et al. (2009) Assessing the impact of transgenerational epigenetic variation on complex traits. *PLoS Genetics,* 5, e1000530.

Khoury, M. J. et al. (2009) Genome-wide association studies, field synopses, and the development of the knowledge base on genetic variation and human diseases. *American Journal of Epidemiology,* 170, pp. 269–279.

Kruglyak, L. (2008) The road to genome-wide association studies. *Nature Review of Genetics,* 9, pp. 314–318.

Kwabi-Addo, B. et al. (2010) Identification of differentially methylated genes in normal prostate tissues from African American and Caucasian men. *Clinical Cancer Research,* 16, pp. 3539–3547.

Lawrence, R. W. et al. (2005) Prospects and pitfalls in whole genome association studies. *Philosophical Transaction of the Royal Society of London B: Biological Sciences,* 360, pp. 1589–1595.

Martinez-Marignac V. L. et al. (2007) Admixture in Mexico City: Implications for admixture mapping of type 2 diabetes genetic risk factors. *Human Genetics,* 120, pp. 807–819.

McKeigue, P. M. (1997) Mapping genes underlying ethnic differences in disease risk by linkage disequilibrium in recently admixed populations. *American Journal of Human Genetics,* 60, pp. 188–196.

———. (1998) Mapping genes that underlie ethnic differences in disease risk: Methods for detecting linkage in admixed populations, by conditioning on parental admixture. *American Journal of Human Genetics,* 63, pp. 241–251.

McKeigue, P. M. et al. (2000) Estimation of admixture and detection of linkage in admixed populations by a Bayesian approach: Application to African-American populations. *Annals of Human Genetics,* 64, pp. 171–186.

Menzel, S. et al. (2007) A QTL influencing F cell production maps to a gene encoding a zinc-finger protein on chromosome 2p15. *Nature Genetics,* 39, pp. 1197–1199.

Montana, G. and Pritchard, J. K. (2004) Statistical tests for admixture mapping with case-control and cases-only data. *American Journal of Human Genetics,* 75, pp. 771–789.

Nei, M. and Roychoudhury, A. K. (1972) Gene differences between Caucasian, Negro, and Japanese populations. *Science,* 177, pp. 434–436.

———. (1974) Genic variation within and between the three major races of man, Caucasoids, Negroids, and Mongoloids. *American Journal of Human Genetics,* 26, pp. 421–443.

Nielsen, D. A. et al. (2010) Ethnic diversity of DNA methylation in the OPRM1 promoter region in lymphocytes of heroin addicts. *Human Genetics,* 127, pp. 639–649.

Novembre, J. et al. (2008) Genes mirror geography within Europe. *Nature,* 456, pp. 98–101.

Paradies, Y. C. et al. (2007) Racialized genetics and the study of complex diseases: The thrifty genotype revisited. *Perspectives in Biology and Medicine,* 50, pp. 203–227.

Parkin, D. M. et al. (2005) Global cancer statistics, 2002. *CA: A Cancer Journal for Clinicians,* 55, pp. 74–108.

Patterson, N. et al. (2004) Methods for high-density admixture mapping of disease genes. *American Journal of Human Genetics,* 74, pp. 979–1000.

Patterson, N. et al. (2006) Population structure and eigenanalysis. *PLoS Genetics,* 2, e190.

Perrine, S. P. (2005) Fetal globin induction—can it cure beta thalassemia? *Hematology (American Society of Hematology Educational Program Book),* pp. 38–44.

Pickrell, J. K. et al. (2009) Signals of recent positive selection in a worldwide sample of human populations. *Genome Research,* 19, pp. 826–837.

Pisani, P. et al. (2002) Estimates of the world-wide prevalence of cancer for 25 sites in the adult population. *International Journal of Cancer,* 97, pp. 72–81.

Pritchard, J. K. et al. (2000) Inference of population structure using multilocus genotype data. *Genetics,* 155, pp. 945–959.

Reich, D. et al. (2005) A whole-genome admixture scan finds a candidate locus for multiple sclerosis susceptibility. *Nature Genetics,* 37, pp. 1113–1118.

Reich, D. et al. (2007) Admixture mapping of an allele affecting interleukin 6 soluble receptor and interleukin 6 levels. *American Journal of Human Genetics,* 80, pp. 716–726.

Reiner, A. P. et al. (2007) Genetic ancestry, population sub-structure, and cardiovascular disease-related traits among African-American participants in the CARDIA Study. *Human Genetics,* 121, pp. 565–575.

Risch, N. (1990) Linkage strategies for genetically complex traits. I. Multilocus models. *American Journal of Human Genetics,* 46, pp. 222–228.

Rosenberg, N. A. et al. (2002) Genetic structure of human populations. *Science,* 298, pp. 2381–2385.

Rosenberg, N. A. et al. (2005) Clines, clusters, and the effect of study design on the inference of human population structure. *PLoS Genetics,* 1, e70.

Rosenberg, N. A. et al. (2010) Genome-wide association studies in diverse populations. *Nature Revue of Genetics,* 11, pp. 356–366.

Ruiz-Narvaez, E. A. et al. (2010) West African and Amerindian ancestry and risk of myocardial infarction and metabolic syndrome in the Central Valley population of Costa Rica. *Human Genetics,* 127, pp. 629–638.

Salari, K. and Burchard, E. G. (2007) Latino populations: A unique opportunity for epidemiological research of asthma. *Paediatric and Perinatal Epidemiology,* 21, Suppl. 3, pp. 15–22.

Sans, M. (2000) Admixture studies in Latin America: From the 20th to the 21st century. *Human Biology,* 72, pp. 155–177.

Sawcer, S. et al. (2010) What role for genetics in the prediction of multiple sclerosis? *Annals of Neurology,* 67, pp. 3–10.

Scherer, M. L. et al. (2010). Admixture mapping of ankle-arm index: Identification of a candidate locus associated with peripheral arterial disease. *Journal of Medical Genetics,* 47, pp. 1–7.

Schuster-Bockler, B. et al. (2010) Dosage sensitivity shapes the evolution of copy-number varied regions. *PLoS One,* 5, e9474.

Seldin, M. F. et al. (2008) Amerindian ancestry in Argentina is associated with increased risk for systemic lupus erythematosus. *Genes and Immunity,* 9, pp. 389–393.

Serre, D. and Pääbo, S. (2004) Evidence for gradients of Human genetic diversity within and among continents. *Genome Research,* 14, pp. 1679–1685.

Shriver, M. D. et al. (2003) Skin pigmentation, biogeographical ancestry and admixture mapping. *Human Genetics,* 112, pp. 387–399.

Sinha, H. et al. (2006) Complex genetic interactions in a quantitative trait locus. *PLoS Genetics,* 2, e13.

Smith, M. W. and O'Brien, S. J. (2005) Mapping by admixture linkage disequilibrium: Advances, limitations and guidelines. *Nature Reviews Genetics,* 6, pp. 623–632.

Smith, E. N. and Kruglyak, L. (2008) Gene-environment interaction in yeast gene expression. *PLoS Biology,* 6, e83.

Stephens, J. C. et al. (1994) Mapping by admixture linkage disequilibrium in human populations: Limits and guidelines. *American Journal of Human Genetics,* 55, pp. 809–824.

Tang, H. et al. (2007) Recent genetic selection in the ancestral admixture of Puerto Ricans. *American Journal of Human Genetics,* 81, pp. 626–633.

Tang, H. et al. (2005) Estimation of individual admixture: Analytical and study design considerations. *Genet Epidemiology,* 28, pp. 289–301.

Tang, H. et al. (2006) Reconstructing genetic ancestry blocks in admixed individuals. *American Journal of Human Genetics,* 79, pp. 1–12.

Tsai, H. J. et al. (2005) Comparison of three methods to estimate genetic ancestry and control for stratification in genetic association studies among admixed populations. *Human Genetics,* 118, pp. 424–433.

Wang, S. et al. (2008) Geographic patterns of genome admixture in Latin American Mestizos. *PLoS Genetics,* 4, e1000037.

Wassel Fyr, C. L. et al. (2007) Genetic admixture, adipocytokines, and adiposity in Black Americans: The Health, Aging, and Body Composition study. *Human Genetics,* 121, pp. 615–624.

Williams, R. C. et al. (2000) Individual estimates of European genetic admixture associated with lower body-mass index, plasma glucose, and prevalence of type 2 diabetes in Pima Indians. *American Journal of Human Genetics,* 66, pp. 527–538.

Zaghloul, N. A. et al. (2010) Functional analyses of variants reveal a significant role for dominant negative and common alleles in oligogenic Bardet-Biedl syndrome. *Proceedings of the National Academy of Sciences (U.S.A.)*, 107, pp. 10602–10607.

Zhu, X. et al. (2005) Admixture mapping for hypertension loci with genome-scan markers. *Nature Genetics*, 37, pp. 177–181.

Zhu, X. et al. (2008) Admixture mapping and the role of population structure for localizing disease genes. *Advances in Genetics*, 60, pp. 547–569.

Ziv, E. et al. (2006) Genetic ancestry and risk factors for breast cancer among Latinas in the San Francisco Bay Area. *Cancer Epidemiology, Biomarkers and Prevention*, 15, pp. 1878–1885.

5

Pharmacogenetics in the Brazilian Population

Guilherme Suarez-Kurtz

Introduction

"Pharmacogenetics deals with pharmacological responses and their modification by hereditary influences."[1] This definition, offered by Werner Kalow in the first book dedicated to pharmacogenetics (Kalow, 1962), highlights the three pillars of this discipline: pharmacology, genetics, and human diversity. Pharmacogenetics has evolved greatly over the 50 years elapsed since Kalow's book was published and was rechristened as pharmacogenomics in the fashion of the "omics" revolution, but its conceptual development and praxis remain contingent upon a better understanding of human genomic diversity and its impact on drug pharmacokinetics and pharmacodynamics. In this chapter, I present an overview of pharmacogenetic/pharmacogenomic (PGx) studies in Brazilians, starting with a brief review of the heterogeneity and structure of the Brazilian population, which have important implications for the conceptual development and clinical implementation of PGx in this country.

Brazil is the fifth-largest country in the world and occupies an area of 8.5 million square kilometers. Its present population exceeds 190 million people, who speak Portuguese, in contrast to all other Latin American nations in which Spanish is the official language. The language reflects the colonization of Brazil by the Portuguese, initiated in the year 1500. At that time, the indigenous population living in the area of what is now Brazil as estimated at approximately 2.5 million (Ribeiro, 1995). The Portuguese-Amerindian admixture started soon after the arrival of the first colonizers. The slave trade of sub-Saharan Africans, from the middle of the sixteenth century until the late 1800s, provided the third major ancestral root of the Brazilian population. Centuries of admixture of Europeans—mainly Portuguese, but also Spaniards, Italians, and Germans—Amerindians, and Africans account for the heterogeneity and diversity of the present-day population of Brazil. The extent of admixture is well documented in a wealth of population genetic studies in Brazilians (Salzano and Bortolini, 2002; Parra et al., 2003; Pena et al. 2009).

In a recent study, G. Suarez-Kurtz et al. (2007a) investigated the genetic characteristics of a cohort of 300 healthy, unrelated individuals living in the city of Rio de Janeiro, in the southeast region of Brazil. Individual DNA was genotyped with a panel of autosomal insertion/deletion (indel) polymorphisms, validated as ancestry-informative markers (Bastos-Rodrigues et al., 2006), and the individual proportions of European, Amerindian, and African biogeographical ancestry were estimated using the software STRUCTURE (Pritchard et al., 2000). Each individual self-identified as white, brown, or black according to the "race/color" classification adopted by the Brazilian census, which relies on self-perception of skin Color. The term "Color" is capitalized to call attention to its special meaning in the context of the Brazilian census classification, Color (in Portuguese, "cor") denoting the Brazilian equivalent of the English term "race."

G. Suarez-Kurtz et al. (2007a) found that individual proportions of Amerindian, European, and African ancestry vary widely among Brazilians, and that most individuals have significant degrees of European and African ancestry, while a sizeable number display Amerindian ancestry also. The average proportions of European ancestry decreased progressively from self-reported white (0.86, $n = 100$), to brown (0.68, $n = 100$), and then to black individuals (0.43, $n = 100$), whereas the opposite trend was observed with respect to African ancestry, which averaged 0.07, 0.24, and 0.50 in white, brown, and black persons, respectively. Amerindian ancestry was relatively constant across the three Color groups, ranging from 0.07 and 0.09. Further analysis of these data is beyond the scope of this review, and herein I will focus on two features that have important implications for PGx: first, European and African components together account for 0.93 (Standard deviation [SD] 0.08) of the genetic diversity in this cohort, as compared to 0.07 (0.08) for Amerindian ancestry. Consequently, European and African ancestry will have a considerably greater impact on the frequency of polymorphisms of PGx relevance, as compared to Amerindian ancestry. Second, the individual proportions of European and African ancestry vary over wide ranges, and most importantly, as a continuum across the Color categories of the Brazilian census. Based on these two features, it is reasonable to anticipate that (1) the greater the difference in frequency of a given polymorphism between

European and sub-Saharan African ancestral populations, the more likely the allele frequency will vary within the African European admixed Brazilian population, and (2) polymorphism frequency will vary continuously as a function of the individual proportions of European and/or African ancestry. These predictions were verified for polymorphisms in several pharmacogenes (CYP3A5, CYP2C8, CYP2C9, CYP2C19, GSTM1, GSTT1, GSTM3, ABCB1, GNB3, and VKORC1), initially in individuals from Rio de Janeiro (Suarez-Kurtz et al., 2007a, 2007b; Estrela et al., 2008a; Vargens et al., 2008), and more recently in a representative cohort of the overall Brazilian population, which included 1,100 self-identified white, brown, and black adults recruited at four different geographical regions (Suarez-Kurtz et al., 2010a, 2010b).

Assessment of the distribution of PGx polymorphisms according to biogeographical ancestry is a novel approach, which I consider more appropriate for Brazilians, and most likely, other admixed populations of the Americas. Nevertheless, most PGx studies in Brazilians are designed, analyzed, and reported according to racial/ethnic categories defined by different criteria and designated by a variety of labels. For example, the terms white, Caucasian, Caucasoid, European derived, Euro Brazilians, or "of European descent" have been used to designate individuals who would, for the most part, self-identify as white according to the Brazilian census. A similar situation is observed with respect to the brown (reported as interethnic admixed, mullato, intermediate, of mixed ancestry, etc.), and black (labeled as black, African derived, Afro- or African-Brazilians, etc.) Color categories. The poor correlation between Color and ancestry in Brazilians casts a shadow of uncertainty over studies that have used only self-reported or investigator assessment of race/color without genomic ancestry analysis. We are fully aware of this caveat, and by adopting the census categorization in our studies we do not expect to circumvent it. Quite on the contrary, we have recently shown that the frequency distribution of CYP2C8 and CYP2C9 polymorphisms among self-identified white, brown, and black Brazilians varies significantly across different geographical regions (Suarez-Kurtz et al., 2010b). Thus, it makes little sense to extrapolate data on CYP2C8 and CYP2C9 polymorphisms from one or more Color strata recruited at a given geographical region (or city) to the ensemble of Brazilians.

PGx Studies in Brazilians

The population diversity of Brazil implies that extrapolation of data derived from relatively well-defined ethnic groups is clearly not applicable to the majority of Brazilians (Suarez-Kurtz, 2008a, 2008b). Only recently, recognition of this fact translated into PGx research on the clinical response to prescribed drugs. By comparison, much more information has accumulated over the last 15 years on genetic variation in metabolic pathways for environmental procarcinogens and its impact on cancer risk in Brazilians. Furthermore, various PGx targets such as alpha- and beta-adrenergic receptors, dopamine and 5HT receptors, components of the renin-angiotensin system, vascular endothelial growth factor, and methylenetetrahydrofolate reductase have been the object of studies of disease susceptibility and phenotypes, rather than drug response in a clinical setting (Suarez-Kurtz and Pena, 2007). Nevertheless, a few academic groups have conducted important PGx research on different therapeutic classes in the Brazilian population. Space limitations do not allow for a comprehensive review (and referencing) of all published PGx clinical trials in Brazilians, but I would like to highlight the contribution of a few distinct groups in this area, before presenting an overview of our own studies. Mara Hutz et al. (Universidade Federal do Rio Grande do Sul) investigated the impact of genetic polymorphisms on the efficacy and toxicity of HMG-CoA reductase inhibitors ("statins") in hypercholesterolemia (Hutz and Fiegenbaum, 2008), on the effects of methylphenidate in attention-deficit/hyperactivity disorders (Kieling et al., 2010), and on the efficacy of clozapine in schizophrenic patients (Kohlrausch et al., 2008). Rosario Hirata et al. (Universidade de São Paulo, USP) explored genetic determinants of the lipid lowering effect of atorvastatin (Rodrigues et al., 2007) and the impact of statins on the expression of ABC drug transporters and CYP3A drug metabolizing enzymes (Rebecchi et al., 2009). José-Eduardo Tanus-Santos (USP) examined the influence of polymorphisms in the endothelial nitric oxide synthase gene on the therapeutic drug response in gestational hypertension and preeclampsia (Sandrim et al., 2010) and on the atorvastatin-induced changes in blood nitrite levels and erythrocyte membrane fluidity (Nagassaki et al., 2009).

Pharmacokinetics of Antiretrovirals

The Brazilian public health system pioneered, in the early 1990s, the universal access to HIV medications free of charge to all citizens who need it. Ritonavir-boosted lopinavir formulations are frequently included in highly active antiretroviral therapy (HAART) regimens for HIV infection. Both drugs are HIV protease inhibitors, but ritonavir at the doses used in co-formulations with lopinavir, serves the purpose of inhibiting CYP3A drug metabolizing enzymes, thereby increasing ("boosting") the exposure to lopinavir, a CYP3A substrate. We studied the impact of polymorphisms in the ABCB1, CYP3A5, and SLCO1B1 genes on the pharmacokinetics of lopinavir and ritonavir in HIV-infected men under stable treatment with Kaletra® at the Hospital Universitário Clementino Fraga Filho, Universidade Federal do Rio de Janeiro. With respect to ABCB1, the study cohort was genotyped for the 1236C > T, 2677G > T/A, and 3435C > T Single Nucleotide Polymorphisms (SNPs). Multivariate regression analysis was applied to assess the influence of ABCB1 geno-types and haplotypes on the trough (predose) concentrations of lopinavir and ritonavir in blood plasma, semen, and saliva (Estrela et al., 2009). We observed marked interindividual variability in the concentrations of both protease inhibitors in the three matrices (e.g., a 27.5-fold range for lopina-vir in plasma) but no association with ABCB1 genotypes or haplotypes.

The goal of the CYP3A5 study in HIV-infected men (Estrela et al., 2008b) was to examine whether the reported impact of the CYP3A5 genotype on the disposition of the HIV protease inhibitor, saquina-vir, administered as a single dose and with no ritonavir boosting to Tanzanian healthy individuals (Josephson et al., 2007) prevailed under a natural clinical setting. Our data revealed no influence of the CYP3A5 genotype on the trough plasma concentration of lopinavir and ritona-vir in HIV-infected males on a stable HAART regimen. We ascribed the apparent discrepancy between our results (Estrela et al., 2008b) and those of F. Josephson et al. (2007) to ritonavir boosting, which was thought to occlude any pharmacokinetic consequences of functional polymorphisms in the CYP3A5 gene. Together, the two studies describe complementary steps toward the implementation of PGx in clinical practice, namely the discovery of association(s) between genetic polymorphisms and phar-macological traits under specific experimental conditions, followed by examination of the impact of the association(s) in the targeted patient populations, studied in the relevant clinical setting.

HIV protease inhibitors have been recently reported to be substrates of the SLCO1B1/OATP1 drug transporter, and a SNP (521T > C) in the SLCO1B1 gene was associated with plasma levels of lopinavir in HIV-infected individuals (Hartkoorn et al., 2010). The availability of plasma concentration data for lopinavir and ritonavir from our cohort of HIV-infected men under stable HAART, provided the opportunity for seek-ing independent confirmation of these results and for extending these observations in two directions: first, by examining two other SLCO1B1 SNPs (388A > G, 463C > A), and second, by exploring the association of the SLCO1B1 polymorphisms with ritonavir plasma concentra-tions (Kohlrausch et al., 2010). Our results confirmed that carriers of the 521C allele display significantly higher lopinavir (but not ritonavir) plasma concentrations relative to the wild-type TT genotype. There was no significant effect of either 388A > G or 463C > A SNPs on lopina-vir or ritonavir plasma concentrations. Reduced OATP-mediated uptake of lopinavir by the hepatocytes in carriers of the SLCO1B1 521C allele was thought to account for the increased plasma levels of lopinavir. The clinical usefulness of this observation is uncertain at present, in view of both the extensive overlap of the trough concentrations of lopinavir in the plasma across the three 521T > C genotypes and the low prevalence (<5 percent) of the homozygous variant genotype (521CC) in most popula-tions. Further studies to confirm the importance of SLCO1B1 polymor-phisms in lopinavir pharmacokinetics are warranted.

Pharmacokinetics and Pharmacodynamics of NSAIDs

The implementation of regulatory legislation for generic drugs in Brazil, in the late 1990s, provided an impetus for bioequivalence studies. The relatively large number of samples collected from healthy subjects under the rigorously controlled conditions of bioequivalence trials offer a favor-able experimental setting for assessing the influence of genetic poly-morphisms on pharmacokinetic parameters. Our group explored this opportunity with respect to the impact of CYP2C9 polymorphisms on the pharmacokinetics and pharmacodynamics of the nonsteroidal anti-inflammatory drugs (NSAIDs), tenoxicam, and piroxicam (Vianna-Jorge et al., 2004; Perini et al., 2005). These were "gene-candidate" studies, the choice of CYP2C9 being based on previous knowledge that the encoded cytochrome P-450 (CYP) enzyme, CYP2C9, provides the major metabolic pathway for the inactivation of both piroxicam and tenoxicam. The poly-morphisms investigated, CYP2C9*2 and CYP2C9*3, encode CYP2C9 isoenzymes with reduced metabolic activity toward these NSAIDs.

The CYP2C9 genotypes had no influence on the peak concentration of either drug in plasma (Cmax). By contrast, the plasma concentration-time curve from zero to infinity (AUCinf), a measure of drug exposure, was significantly greater in CYP2C9*1/*2 or *1/*3 heterozygous, as com-pared to wild-type homozygous subjects (CYP2C9*1/*1). This effect was associated with and explained by the significantly lower oral clearance (corrected for body weight, CL/Fcor) of either NSAID in heterozygous carriers of the CYP2C9*2 or *3 alleles, compared to wild-type homozy-gous subjects.

The piroxicam study was extended to include a pharmacodynamic parameter, namely the ex vivo production of thromboxane B2 (TxB2), catalyzed by cyclooxygenase 1 (COX-1), a distinct target of NSAIDs. Our work hypothesis was that inhibition of COX-1 by piroxicam would be greater in carriers of the defective CYP2C9*2 and CYP2C9*3 alleles, because of the increased exposure to piroxicam in these individuals, as compared to wild-type homozygous (Perini et al., 2005). The results verified our hypothesis by showing significant differences in TxB2 production between either CYP2C9*1/*2 or CYP2C9*1/*3 genotypes and CYP2C9*1/*1. The ANOVA R2 (coefficient of variation) among the three genotypes indicated that 78 percent of the interindividual variability may be explained by the CYP2C9 polymorphisms examined, namely CYP2C9*2 and CYP2C9*3. This is a remarkable example of PGx modulation, as there are not many instances of such a large contribution of genetic polymorphisms to the interindividual variability of pharmacodynamic phenotypes.

The piroxicam bioequivalence trial disclosed one individual who displayed a unique pharmacokinetic profile, such that his plasma piroxicam concentration 13 days after the administration of a single drug dose remained at 70 percent of the peak concentration measured at 2 hours (estimated half-life of elimination, 500 hours). Nevertheless, no adverse effects were observed or reported by the individual. After a six-month washout, when piroxicam was no longer detected in his plasma, this individual was reexposed to a single piroxicam dose, and blood samples were collected for 120 days for CYP2C9 genotyping, measurements of piroxicam plasma concentrations and assessment of the activity of COX-1 and COX-2—the pharmacodynamic targets of NSAIDs—through the ex vivo formation of TBX2 and prostaglandin E2, respectively (Perini and Suarez-Kurtz, 2006). The individual was genotyped as CYP2C9*3/*3, an extremely rare genotype in Brazilians, which encodes a nonfunctional CYP2C9 isoform. The impact of this genotype was remarkable: the effects on pharmacokinetic parameters observed with the first dose were reproduced, and the pharmacodynamic analyses showed that COX-2 activity was maintained below 50 percent of its control value for more than one month and that COX-1 was below 20 percent of its control value for more than two months after the single piroxicam dose! Again, no adverse effects were observed. We speculated on the potential advantages conferred by the PGx polymorphisms associated with impaired drug metabolism for long-term strategies in disease prevention and treatment (Perini and Suarez-Kurtz, 2006).

Warfarin Dosing Algorithms for Brazilians

Warfarin combines several characteristics that make it a model target for individualized drug therapy: (1) it is the most commonly prescribed oral anticoagulant in Brazil and worldwide, (2) there is large interindividual variation in the required dosage, (3) its therapeutic index is narrow, and incorrect dosage, especially during the initial phase of treatment, carries a high risk of bleeding or failure to prevent thromboembolism, and (4) a reliable biomarker, the international normalized ratio (INR), is available for quantifying warfarin's anticoagulant effect. Polymorphisms in two genes, namely CYP2C9 and VKORC1, have been repeatedly found to associate with the clinical response to warfarin. CYP2C9 encodes CYP2C9, the primary metabolic route for the disposition of S-warfarin, the most active isomer of clinically used racemic warfarin. The polymorphic VKORC1 gene codifies VKORC1, the molecular target of coumarin anticoagulants. In collaboration with the Instituto Nacional de Cardiologia Laranjeiras (INCL), a reference cardiology hospital of the Brazilian public health system located in Rio de Janeiro, we assessed the contribution of CYP2C9 and VKORC1 polymorphisms and of demographic and clinical variables to the interindividual variability in warfarin dose requirement for stable anticoagulation, defined as three successive INR readings in the 2–3.5 range (Perini et al., 2008a). The warfarin dose required for stable anticoagulation varied from 3 to 75 mg/week, that is, a 25-fold interindividual variability! Multivariate regression modeling was applied to develop a dosing algorithm for prediction of the appropriate warfarin dose for each patient. The VKORC1 3673G > A genotype proved to be the most important predictor in our model, with a partial R^2 value of 23.8 percent, followed by the number of copies of variant CYP2C9 alleles (R^2 = 6.9 percent) and by co-treatment with amiodarone (R^2 = 5.6 percent) (Perini et al. 2008a, 2008b). Body weight, therapeutic indication for the use of warfarin, age, and co-treatment with simvastatin were the other covariates associated with warfarin dose requirement. A dosing algorithm including all these covariates explained 51 percent of the variance in warfarin dose requirement in our study cohort.

We extended our studies on the INCL cohort to explore the association of other variables with warfarin dose requirements (Suarez-Kurtz et al., 2009) and reported that the inclusion of INR/dose as a covariate in regression modeling of the stable warfarin dose leads to a novel algorithm with greater predictive power (R^2 = 60 percent). The most informative model retained not only the same covariates previously identified as associated with stable warfarin weekly dose in this cohort (age, weight, treatment indication, co-medication with amiodarone, or simvastatin, VKORC1, and CYP2C9 genotypes) but included also an INR/dose term. A distinct feature of the novel algorithm is that the individual INR/dose term does not represent a fixed time point after starting warfarin therapy, but rather the first measurement taken after admission of the patients in the anticoagulant unit. This feature is potentially useful for patients under continuous warfarin treatment, who had not reached stable dosing despite repeated dose adjustments.

In a recently published study (Perini et al., 2010), we investigated the association between warfarin dose requirement and a polymorphism in CYP4F2, namely, rs2108622 C > T (V33M). Biological plausibility for this association was provided by the evidence that CYP4F2 is a vitamin K1 oxidase and carriers of the CYP4F2 33M allele are likely to have elevated hepatic levels of vitamin K1, requiring higher warfarin doses (McDonald et al., 2009). Previous studies had reported controversial results regarding the contribution of the CYP4F2 rs2108622 to the warfarin dose requirement for stable anticoagulation but had one common feature: the enrolled patients were exclusively or predominantly (>90 percent) white, of European descent. The allele frequency of rs2108622 differs markedly among populations, being lower in sub-Saharan Africans and African Americans (0–9 percent) as compared with Europeans and white North Americans (17–33 percent). From a population perspective, it might be anticipated that rs2108622 will have a smaller influence, if any, on the warfarin dose requirements in black individuals. We explored this hypothesis in the INCL cohort, which included self-identified white, brown, and black individuals, and observed that addition of the rs2108622 genotype as a variable had only a marginal effect on the predictive power of the warfarin dosing algorithm previously derived from this patient cohort. We concluded that prospective CYP4F2 genotyping is not justified in Brazilians who are potential candidates for warfarin therapy (Perini et al., 2010).

Collectively, out studies on the PGx of warfarin in Brazilians have distinct strengths. First, they reflect real-life community prescribing and dispensing of warfarin in the context of a public hospital in a developing country. Second, the recruitment of patients from the notoriously admixed and heterogeneous Brazilian population allowed for the investigation of PGx associations in individuals with heterogeneous genetic ancestry, under the same environmental conditions. A distinct feature of a PGx algorithms derived from our patient cohort is the ability to predict equally well the warfarin dose requirement in white and black patients, in marked contrast with the poorer outcome of several warfarin dosing algorithms in blacks, notably African Americans, as compared to white patients (Gage et al., 2008). Third, the nongenetic covariates of our dosing algorithms are readily available in the patient medical record, whereas the genetic covariates, namely the CYP2C9 and VKORC1 3673G > A genotypes, can be determined in a time frame compatible with the urgency of starting warfarin administration to many patients. However, we also recognize that our studies have limitations, most of which are common to studies of similar design performed in other populations. First, by being retrospective, the associations and dosing algorithm described in our studies must be prospectively tested in an appropriately large cohort. Second, we did not assess the impact of the algorithms on the clinical response to warfarin but instead used the INR as a surrogate for it. Third, by explaining 51–60 percent of the variance in warfarin dosing, our algorithms leave 40–50 percent of this variance to be accounted for by other variables, whether genetic or nongenetic. Finally, our algorithms, developed for a heterogeneous and admixed population of a large Brazilian city, may not be applicable to other population groups.

Studies in Amerindian Groups Living in Brazil

Native American (Amerindian) populations are poorly represented in PGx research and PGx databases (reviewed by Jaja et al., 2008). The extant Amerindian population of Brazil is very diverse, consisting of more than 227 ethnic groups speaking more than 180 languages. This diversity is reflected in the frequency of polymorphisms in genes encoding drug metabolizing enzymes and drug targets (reviewed in Suarez-Kurtz and Pena, 2007). In collaboration with Maria Luiza Petzl-Erler et al. from Universidade Federal do Paraná, we assessed the distribution of four SNPs in VKORC1 in three endogamous Amerindian populations, namely Guarani-Kaiowá, Guarani-Ñandeva, and Kaingang, living in indigenous reservation areas in the south and center-west regions of Brazil (Perini et al., 2008b). Despite living side by side for centuries, the Guarani (Tupi-Guarani linguistic family) and the Kaingang (Gê-speaking people) maintain their cultural and linguistic distinctiveness, which is paralleled by substantial genetic distance between them. Our results indicated that the VKORC1 5808T > G SNP was absent or rare (<3 percent), whereas 3673G > A, 6853G > C, and 9041G > A were frequent (34–63 percent) in the three Amerindian populations. No difference was detected in VKORC1 allele or genotype frequency between the two Guarani populations, whereas significant differences were observed between Kaingang and Guarani. The data disclosed a uniqueness of the frequency distribution of the VKORC1 SNPs in the Guarani and Kaingang, compared to Asian, African, and European populations. However, in view of the vast interpopulational diversity among Amerindians, these results should not be interpreted as representative of other extant Amerindian peoples. Our observation that 40 percent of Kaingang and 60 percent of Guarani have VKORC1 haplotypes that include the variant 3673A allele associated with increased susceptibility to warfarin suggests that these two Amerindian populations comprise high proportions of individuals requiring reduced warfarin doses for appropriate anticoagulation in the prevention or treatment of thromboembolism.

Studies in Brazilians with Japanese Ancestry

Brazil hosts the largest Japanese community outside Japan, estimated at 1.5 million individuals, one third of whom are first generation, Brazilian born with native Japanese parents. This large community provided a unique opportunity for comparative studies of the distribution of PGx polymorphisms in native Japanese versus their Brazilian-born descendants. In collaboration with ândrea Ribeiro-dos-Santos et al. (Universidade Federal do Pará) and Emilio Moriguchi (Universidade Federal do Rio Grande do Sul), we explored this opportunity with respect to functional polymorphisms in genes that modulate drug disposition (CYP2C9, CYP2C19, and GSTM3) or response (VKORC1), and that differ significantly in frequency in native Japanese versus Brazilians with no Japanese ancestry (Perini et al., 2009). Our study detected no difference in the frequency of the CYP2C9*2, CYP2C9*3, CYP2C19*2, CYP2C19*3, GSTM3*A/*B, or VKORC1 3673G > A, 5808T > G, 6853G > C, and 9041G > A polymorphisms between native Japanese and first-generation Japanese descendants. In contrast, significant differences in the frequency of each polymorphism were observed between native or first-generation Japanese and Brazilians with no Japanese ancestry. The striking similarity in the frequency of clinically relevant PGx polymorphisms between Brazilian-born Japanese descendants and native Japanese led us to suggest that the former may be recruited for clinical trials designed to generate bridging data for the Japanese population in the context of the International Conference on Harmonization for drug development.

The Brazilian Pharmacogenetic Network (Refargen)

To deal with the specificity of PGx in Brazilians, a nationwide network named Rede Nacional de Farmacogenética (Brazilian National Pharmacogenetics Network, Refargen) was established in 2004 (Suarez-Kurtz, 2004). Presently, the network congregates 18 research groups from various Brazilian institutions, and its mission is to provide leadership in PGx research and education in Brazil, with a population health impact. Information on Refargen's members and activities, as well as an updated database of allele, genotype, and haplotype frequencies of PGx polymorphisms in the Brazilian population may be accessed at the network Internet site.

Concluding Remarks

Personalized drug therapy, the promise of pharmacogenomics, must be based on the recognition of the inherent genetic individuality. This notion is particularly relevant to admixed populations, in which substructure increases further the fluidity of racial/ethnic labels (Suarez-Kurtz, 2005). Because interethnic admixture is either common or increasing at a fast pace in many, if not most populations, extrapolation on a global scale of PGx data from well-defined ethnic groups is plagued with uncertainty. To impact positively on global health, PGx must broaden its scope, with respect to both target and population diversity, and be inclusive of admixed populations with their perceived challenges and advantages. From this perspective, PGx studies in Brazilians have the potential to contribute relevant information toward personalized drug prescription worldwide.

Acknowledgments

Research in the author's laboratory is sponsored by Conselho Nacional de Desenvolvimento Científico e Tecnológico (CNPq), Coordenação de Aperfeiçoamento de Pessoal de Nivel Superior (CAPES), Financiadora de Estudos e Projetos (Finep), and Fundação de Amparo à Pesquisa do Estado do Rio de Janeiro (Faperj).

Notes

First published in *Frontiers in Pharmacology*, 1, pp. 1–10, 2010

1. This chapter is an abridged version of the paper by Suarez-Kurtz (2010).

References

Bastos-Rodrigues, L. et al. (2006) The genetic structure of human populations studied through short insertion-deletion polymorphisms. *Annals of Human Genetics*, 70, pp. 658–665.

Estrela, R. C. E. et al. (2008a) The distribution of ABCB1 polymorphisms among Brazilians: Impact of population admixture. *Pharmacogenomics*, 9, pp. 267–276.

Estrela, R. C. E. et al. (2008b) CYP3A5 genotype has no impact on plasma trough concentrations of lopinavir and ritonavir in HIV-infected subjects. *Clinical Pharmacology & Therapeutics*, 84, pp. 205–207.

Estrela, R. C. E. et al. (2009) ABCB1 polymorphisms have no impact on the concentrations of lopinavir and ritonavir in blood, semen and saliva of HIV-infected men under stable antiretroviral therapy. *Pharmacogenomics*, 10, pp. 311–318.

Gage, B. F. et al. (2008) Use of pharmacogenetic and clinical factors to predict the therapeutic dose of warfarin. *Clinical Pharmacology & Therapeutics*, 84, pp. 326–331.

Hartkoorn, R. C. et al. (2010) HIV protease inhibitors are substrates for OATP1A2, OATP1B1 and OATP1B3 and lopinavir plasma concentrations are influenced by SLCO1B1 polymorphisms. Pharmacogenet. *Genomics*, 20, pp. 112–120.

Hutz, M. H., and Fiegenbaum, M. (2008) Impact of genetic polymorphisms on the efficacy of HMG-CoA reductase inhibitors. *American Journal of Cardiovascular Drugs*, 8, pp. 161–170.

Jaja, C. et al. (2008) Cytochrome p450 enzyme polymorphism frequency in indigenous and native American populations, a systematic review. *Community Genetics*, 11, pp. 141–149.

Josephson, F. et al. (2007) CYP3A5 genotype has an impact on the metabolism of the HIV protease inhibitor saquinavir. *Clinical Pharmacology & Therapeutics,* 81, pp. 708–712.

Kalow, W. (1962) *Pharmacogenetics, Heredity and the Response to Drugs.* Philadelphia: WB Saunders Co.

Kieling, C. et al. (2010) A current update on ADHD pharmacogenomics. *Pharmacogenomics*, 11, pp. 407–419.

Kohlrausch, F. B. et al. (2010) The impact of SLCO1B1 polymorphisms on the plasma concentration of lopinavir and ritonavir in HIV-infected men. *British Journal of Clinical Pharmacology*, 69, pp. 95–98.

Kohlrausch, F. B. et al. (2008) G-protein gene 825C > T polymorphism is associated with response to clozapine in Brazilian schizophrenics. *Pharmacogenomics*, 9, pp. 1429–1436.

McDonald, M. G. et al. (2009) CYP4F2 is a vitamin K1 oxidase: An explanation for altered warfarin dose in carriers of the V433M variant. *Molecular Pharmacology*, 75, pp. 1337–1346.

Nagassaki, S. et al. (2009) eNOS T-786C polymorphism affects atorvastatin-induced changes in erythrocyte membrane fluidity. *European Journal of Clinical Pharmacology*, 65, pp. 385–392.

Parra, F. C. et al. (2003) Color and genomic ancestry in Brazilians. *Proceedings of the National Academy of Sciences of the United States of America*, 100, pp. 177–182.

Pena, S. D. J. et al. (2009) DNA tests probe the genomic ancestry of Brazilians. *Brazilian Journal of Medical and Biological Research*, 42, pp. 870–876.

Perini, J. A. et al. (2008a) Pharmacogenetics of warfarin: Development of a dosing algorithm for Brazilian patients. *Clinical Pharmacology & Therapeutics,* 84, pp. 722–728.

Perini, J. A. et al. (2008b) VKORC1 polymorphisms in Amerindian populations of Brazil. *Pharmacogenomics*, 9, pp. 1623–1629.

Perini, J. A. et al. (2010) Impact of CYP4F2 rs2108622 on the stable warfarin dose in an admixed patient cohort. *Clinical Pharmacology & Therapeutics*, 87, pp. 417–420.

Perini, J. A., and Suarez-Kurtz, G. (2006) Impact of CYP2C9*3/*3 genotype on the pharmacokinetics and pharmacodynamics of piroxicam. *Clinical Pharmacology & Therapeutics*, 80, pp. 549–551.

Perini, J. A. et al. (2009) Pharmacogenetic polymorphisms in Brazilian-born, first-generation Japanese descendants. *Brazilian Journal of Medical and Biological Research*, 42, pp. 1179–1184.

Perini, J. A. et al. (2005) Influence of CYP2C9 genotypes on the pharmacokinetics and pharmacodynamics of piroxicam. *Clinical Pharmacology & Therapeutics*, 78, pp. 362–369.

Pritchard, J. K. et al. (2000) Inference of population structure using multilocus genotype data. *Genetics*, 155, 945–959.

Rebecchi, I. M. et al. (2009) ABCB1 and ABCC1 expression in peripheral mononuclear cells is influenced by gene polymorphisms and atorvastatin treatment. *Biochemical Pharmacology*, 77, pp. 66–75.

Ribeiro, D. (1995) *O povo brasileiro: A formação e o sentido do Brasil*. São Paulo: Companhia das Letras.

Rodrigues, A. C. et al. (2007) The genetic determinants of atorvastatin response. *Current Opinion in Molecular Therapeutics*, 9, pp. 545–553.

Salzano, F. M., and Bortolini, M. C. (2002) *The Evolution and Genetics of Latin American Populations*. Cambridge: Cambridge University Press.

Sandrim, V. C. et al. (2010) eNOS haplotypes affect the responsiveness to antihypertensive therapy in preeclampsia but not in gestational hypertension. *Pharmacogenomics Journal*, 10, pp. 40–45.

Suarez-Kurtz, G. (2004) Pharmacogenomics in admixed populations: The Brazilian pharmacogenetics/pharmacogenomics network—REFARGEN. *Pharmacogenomics Journal*, 4, pp. 347–438.

Suarez-Kurtz, G. (2005) Pharmacogenetics in admixed populations. *Trends in Pharmacological Sciences*, 26, pp. 196–201.

Suarez-Kurtz, G. (2008a) The implications of population admixture in race-based therapy. *Clinical Pharmacology & Therapeutics*, 83, pp. 399–400.

Suarez-Kurtz, G. (2008b) Ethnic differences in drug therapy, a pharmacogenomics perspective. *Expert Review in Clinical Pharmacology*, 1, pp. 337–339.

Suarez-Kurtz, G. (2010) Pharmacogenetics in the Brazilian population. *Frontiers in Pharmacology*, 1, pp. 1–18.

Suarez-Kurtz, G. et al. (2010a) VKORC1 polymorphisms in Brazilians, comparison with the Portuguese and Portuguese-speaking Africans and pharmacogenetic implications. *Pharmacogenomics*, 11, pp. 1257–1267.

Suarez-Kurtz, G. et al. (2010b) Global pharmacogenomics: Impact of population diversity on the distribution of polymorphisms in the CYP2C cluster among Brazilians. *Pharmacogenomics J*, Dec 21 epub ahead of print.

Suarez-Kurtz, G. and Pena, S. D. J. (2007) Pharmacogenetic studies in the Brazilian population. In: Suarez-Kurtz, G. (ed.) Pharmacogenomics in Admixed Populations. Austin: Landes, pp. 75–98.

Suarez-Kurtz, G. et al. (2007a) Self-reported skin color, genomic ancestry and the distribution of GST polymorphisms. *Pharmacogenetics and Genomics*, 17, pp. 765–771.

Suarez-Kurtz, G. et al. (2007b) Impact of population admixture on the distribution of the CYP3A5*3 polymorphism. *Pharmacogenomics*, 8, pp. 1299–1306.

Suarez-Kurtz, G. et al. (2009) Relative contribution of VKORC1, CYP2C9 and INR response to warfarin stable dose. *Blood*, 113, pp. 4125–4126.

Vargens, D. D. et al. (2008) Distribution of the GNB3 825C > T polymorphism among Brazilians, impact of population structure. *European Journal of Clinical Pharmacology*, 3, pp. 253–256.

Vianna-Jorge, R. et al. (2004) CYP2C9 genotypes and the pharmacokinetics of tenoxicam in Brazilians. *Clinical Pharmacology & Therapeutics*, 76, pp. 18–26.

6

Strong Association of Socioeconomic Status and Genetic Ancestry in Latinos: Implications for Admixture Studies of Type 2 Diabetes

J. C. Florez, A. L. Price, D. Campbell, L. Riba,
M. V. Parra, F. Yu, C. Duque, R. Saxena,
N. Gallego, M. Tello-Ruiz, L. Franco,
M. Rodríguez-Torres, A. Villegas, G. Bedoya,
C. A. Aguilar-Salinas, M. T. Tusié-Luna,
A. Ruiz-Linares, and D. Reich

Introduction

In comparison with the European American population, minority groups in the United States experience a disproportionate burden of type 2 diabetes (Carter et al. 1996). This is particularly evident in some Native American tribes, with the Pima Indians presenting one of the highest population prevalences of type 2 diabetes ever reported for any ethnic group (Bennett et al. 1971; Hamman et al. 1978; Knowler et al. 1990). A similar trend can be observed among the admixed populations of Mexico, the Caribbean, and Central and South America (including recent immigrants to the United States), who are commonly characterized as Latinos. Indeed, the risk of type 2 diabetes is twofold to fivefold higher in Latinos from Puerto Rico, Texas, New Mexico, or Colorado (Stern et al. 1981; Hanis et al. 1983; Samet et al. 1988; Hamman et al. 1989; Flegal et al. 1991) than in whites, an epidemiological difference that persists after adjusting for other traits such as abdominal obesity (Haffner et al. 1986).

These differences support the notion that diabetes risk factors occur at a higher frequency in populations of Native American descent. A genetic component for this bias was suggested in the form of the "thrifty gene" hypothesis and its interaction with the environment first posed by J. Neel over 40 years ago (Neel 1962). A similar hypothesis was proposed for Mexican Americans, based on crude skin pigmentation measures (Gardner et al. 1984) and subsequently supported by analysis of genetic markers (Chakraborty et al. 1986). Initial evidence of a genetic contribution to the higher risk of type 2 diabetes in Native Americans was provided by the observation that Pima Indians with type 2 diabetes have significantly more Native American ancestry than their normoglycemic counterparts (Knowler et al. 1988), and that the fraction of European ancestry was associated with protective metabolic phenotypes in this population (Williams et al. 2000); the potential contribution of socioeconomic status to these estimates was not fully assessed. In addition, the recent identification of the *ABCA1* R230C variant in Mexican admixed individuals (mestizos) illustrates that exclusive gene variants derived from Native American ancestry can indeed influence type 2 diabetes risk (Villarreal-Molina et al. 2007; Villarreal-Molina et al. 2008).

The advent of comprehensive databases cataloging genetic variation (Reich et al. 2003), the development of high-throughput genotyping technologies and the availability of DNA samples from multiple populations make it possible to select a set of single nucleotide polymorphisms (SNPs) that are highly informative for geographic ancestry, commonly termed ancestry-informative markers (AIMs). Thus, SNPs that have widely divergent allele frequencies in ancestral populations can be compiled to make such determinations of ancestral origin. When evenly spaced throughout the genome, approximately 2,000 AIMs can be employed to infer ancestry at each genomic location for admixture mapping, but a much smaller set of randomly distributed AIMs (<100) can also be genotyped in admixed persons to derive a fairly accurate estimate of genetic ancestry, expressed as a proportion of each individual's genome. Such compilations specific to Latino populations have recently been done by three different groups, including our own (Mao et al. 2007; Price et al. 2007; Tian et al. 2007).

In anticipation of genome-wide admixture mapping, AIMs have been applied to Latino samples with the goal of estimating the genetic contribution to the increased diabetes prevalence in this population. An initial study performed in American Latinos and stratified by neighborhood detected a strong association between Native American ancestry and socioeconomic status; however, the authors concluded that despite the presence of such confounding, a genetic component to the increased disease prevalence was likely (Chakraborty et al. 1986). The association between type 2 diabetes and Native American ancestry was further substantiated by Parra et al. in Mexican Americans from the San Luis Valley, Colorado; but once again, controlling for income and education abolished the statistical significance of the finding (Parra et al. 2004). More recently, a similar study was conducted in a sample of 286 unrelated diabetic patients and 275 controls assembled from users of the Social Security hospital in Mexico City, which is thought to capture a large middle segment of the population devoid of upper- and lower-income outliers. This report found a nonsignificant increase in Native American ancestry among participants with diabetes, but a much stronger association between higher educational level and both European ancestry and nondiabetic status (Martinez-Marignac et al. 2007).

The success of admixture mapping in Latino populations is predicated on the ability to dissect these extra-genetic confounders from the genetic association. The correlation of socioeconomic status with ancestry in samples from two distinct US American locations and in an additional sample from Mexico City suggests that this may be a general phenomenon. To replicate and expand these observations, and as a way of investigating its likely impact on ongoing admixture mapping studies, we evaluated the contribution of socioeconomic status to the ancestry-diabetes relationship in two separate, non-US Latino populations from North and South America.

Methods

Populations

We studied 499 patients with type 2 diabetes and 197 controls from Medellín (Colombia), as well as 163 patients with type 2 diabetes and 72 controls from central Mexico. Demographic characteristics are presented in table 6.1. In Colombia, diabetic patients were recruited from diabetes clinics in and around Medellín, with diagnostic criteria including fasting plasma glucose ≥7 mmol/L or 2 h glucose >11.1 mmol/L after a 75 g OGTT. Controls were unaffected participants over 40 years of age who had no history of diabetes among first-degree relatives (and at times also including grandparents). Medellín is divided into main sectors ("comunas"), and the mean socioeconomic status is available for each one. We

Table 6.1   Demographic characteristics of genotyped samples

	Mexico		Colombia	
	Cases	Controls	Cases	Controls
N	163	72	499	197
Men/women	72/91	32/40	183/316	53/144
Age (years)	48.5±12.5	56.0±9.9	64.1±10.3	60.7±9.9
BMI (kg/m²)	28.1±5.8	25.6±3.7	26.7±4.6	25.4±4.1

Notes: Quantitative traits are means±SD

balanced the collection of patients and controls in Medellín by select-
ing the diabetes clinics (for the patients) and the centers for the care of
the elderly (for the controls) from the same sectors of the city, with the
express intention of reflecting a range of socioeconomic status strata. In
Mexico, type 2 diabetic participants included individuals treated at the
Diabetes Clinic of the Instituto Nacional de Ciencias Médicas y Nutrición
in Mexico City, in whom diabetes was confirmed with a blood glucose
sample obtained after a nine- to twelve-hour fast. Exclusion criteria
included secondary causes of diabetes (e.g., diseases of the exocrine pan-
creas, endocrinopathies, both drug or chemical induced), genetic syndromes associated with diabetes and insulin therapy during the first two
years after diagnosis. Controls were selected among spouses of diabetic
participants or unrelated patients who were seeking medical attention at
the same Instituto Nacional de Ciencias Médicas y Nutrición for reasons
other than diabetes (e.g., primary dyslipidemia) and were older than 40
years of age, had no history of diabetes among first-degree relatives, and
had fasting plasma glucose less than 6.1 mmol/L. Individuals identified
themselves as Mexican mestizos with both parents and grandparents
born in Mexico. By this ascertainment strategy we tried to ensure that
our cases and controls were selected from the same geographical area and
had similar socioeconomic status and comparable access to the public
health system.

To estimate allele frequencies in the ancestral populations and project ancestry proportions, we also studied several unmixed populations:
European Americans from Baltimore and Chicago ($n=77$) and from
the HapMap Centre d'Etude du Polymorphisme Humain (Utah residents with northern and western European ancestry) collection ($n=60$);
Spaniards from Valencia ($n=31$); West Africans from Ghana ($n=52$) and
from the HapMap Yoruba in Ibadan, Nigeria collection ($n=60$); and
Native Americans from the Mazahua ($n=22$), Zapotec ($n=60$), Mixtec
($n=23$), and Mixe ($n=29$) populations. All participants gave informed

consent, and studies were carried out in accordance with the principles of the Declaration of Helsinki as revised in 2000.

Estimation of Socioeconomic Status

In Colombia, we had access to a government-assigned "property band" based on property valuation of each individual home for the purposes of billing for public utilities and ranging from one (lowest) to six (highest). An assignment to one of these strata was made for each participant based on the telephone number they provided at the time of their interview. We contrasted this information with other data we collected (such as home or car ownership) and confirmed a close correlation between the governmental water-usage rank and these other markers of socioeconomic status.

In Mexico, socioeconomic status was determined by social workers at the National Institute of Medical Sciences and Nutrition using a standardized and validated tool currently applied in all Mexican National Institutes of Health studies (Silva Arciniega and Brain Calderon 2006). Questionnaires include information on six categories: family monthly income, occupation of the head of the household, percentage of family income spent on food, type and characteristics of residence (owner occupied, rented, or shared with extended family), place of residence, and the presence of chronic illnesses in other family members. Points are given for each category and the sum is used to assign participants to one of six socioeconomic status bands (lowest to highest). Supporting documents are required to validate the information. When information was not complete or questionable, socioeconomic status assignments were further explored and confirmed via unscheduled home visits.

Genotyping

Samples from all populations were genotyped using the Sequenom MassARRAY technology (Tang et al. 1999) at 67 AIMs (Electronic supplementary material [ESM] table 6.1). These SNPs are on an average 49 percent different in frequency between Native Americans and European Americans and are spaced by at least 10 cM (or >10 Mb) on chromosomes 1 to 22 (see supplementary table B in Smith et al. 2004, from which these markers were selected). The large number of markers and the high degree of informativeness per marker in this set yield precise estimates of the proportion of European ancestry for each individual and also a measure of the precision of each estimate as a standard error. The average standard error for the percentage European ancestry estimate was ±7.2 percent (in Mexicans) and ±8.0 percent (in Colombians) respectively. We had complete data for 89 percent of genotypes; this reduction below 100 percent reflects the fact that slightly different panels of markers were genotyped on some samples. The missing data are not expected to affect our estimates of ancestry proportion.

Statistical Methods

To search for genetic differences associated with ancestry between participants with type 2 diabetes and controls (specifically, underrepresentation of European ancestry in diabetic participants), we used a mixture-of-binomials model to estimate the ancestry proportions of each Latino individual as previously described (Price et al. 2007) and compared the distributions of ancestries in cases versus controls (table 6.2). We note that Latino populations represent a three-way admixture of European, Native American, and African ancestry. Thus, we had a choice of whether to base our analyses on percent European ancestry (distinguishing European vs. [Native American + African]) or on percent Native American ancestry (distinguishing Native American vs. [European + African]). Because the 67 AIMs used to infer ancestry were relatively less informative for Native American versus African ancestry, the standard errors for inferred percent European ancestry were smaller than the standard errors for inferred percent Native American ancestry; therefore, we based our analyses on percent European ancestry.

Associations of European ancestry proportion, socioeconomic status, and/or BMI with type 2 diabetes status were assessed via logistic regression with (1) European ancestry proportion, (2) socioeconomic status, (3) socioeconomic status and European ancestry proportion, or

Table 6.2   Association of non-European ancestry with type 2 diabetes in Latinos

Country of origin	n (cases)	n (controls)	European ancestry, cases (%)	European ancestry, controls (%)	Logreg coefficient c_{anc}	Pseudo-r^2 for c_{anc}	p value for c_{anc}
Mexico	163	72	33±20	46±24	−2.84	0.08	2×10^{-5}
Colombia	499	197	56±15	59±14	−1.36	0.01	0.02

Notes: Unless otherwise indicated, values are means±SD
We report average ancestry in cases and controls, as well as results of a logistic regression using ancestry (c_{anc}) as a predictor of disease outcome.

(4) socioeconomic status, European ancestry proportion, and BMI as covariates. We included a constant term in each regression analysis. For example, letting π denote disease outcome, the logistic regression model for (4) was

$$\pi = \frac{e^{c_0 + c_{\text{SES}}\text{SES} + c_{\text{anc}}\text{anc} + c_{\text{BMI}}\text{BMI}}}{1 + e^{c_0 + c_{\text{SES}}\text{SES} + c_{\text{anc}}\text{anc} + c_{\text{BMI}}\text{BMI}}}$$

where c_{BMI} is the logistic regression coefficient multiplying BMI, anc is genome-wide ancestry, c_{anc} is the logistic regression coefficient multiplying anc, SES is socioeconomic status, c_{SES} is the logistic regression coefficient multiplying socioeconomic status, and c_0 is the logistic regression coefficient multiplying the constant term.

For each regression analysis, we computed effect sizes, p values and pseudo-r^2 values. We defined pseudo-r^2 as the reduction in magnitude of the mean square value of disease outcome minus the predicted probability of disease outcome, comparing each of the four logistic regression models to a constant-term only model.

Results

We analyzed the relationship between genetic ancestry and type 2 diabetes status in two Latino populations from Mexico and Colombia. Both of these populations inherit European, Native American, and African ancestry, but we focused our analysis on the proportion of European ancestry as inferred from using 67 AIMs. In the Mexicans, we observed a statistically significant difference (odds ratio [OR] [95% CI] 0.06 [0.02–0.21], $p=2 \times 10^{-5}$) in genetic ancestry between patients with type 2 diabetes and age- and sex-matched controls from central Mexico (table 6.2). A similar phenomenon, although of lesser magnitude, was observed for the Colombians, with a nominally significant difference (OR 0.26 [0.08–0.78], $p=0.02$) in genetic ancestry between diabetic participants and nondiabetic controls (table 6.2). This ancestry difference was due to participants with type 2 diabetes having less European ancestry in both groups (figure 6.1). Overall, the proportion of European ancestry is estimated to be 33 percent in diabetic participants versus 46 percent in controls in Mexicans, and 56 percent versus 59 percent respectively in Colombians.

We investigated whether the observed association between genetic ancestry and type 2 diabetes was confounded by socioeconomic status. We determined that socioeconomic status was strongly correlated with European ancestry proportion in Mexicans (33 percent correlation,

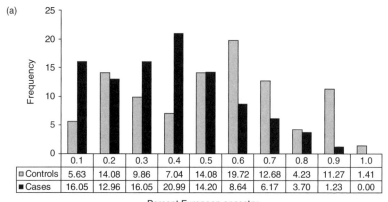

(a)	0.1	0.2	0.3	0.4	0.5	0.6	0.7	0.8	0.9	1.0
Controls	5.63	14.08	9.86	7.04	14.08	19.72	12.68	4.23	11.27	1.41
Cases	16.05	12.96	16.05	20.99	14.20	8.64	6.17	3.70	1.23	0.00

Percent European ancestry

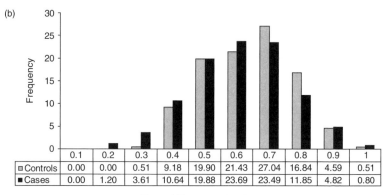

(b)	0.1	0.2	0.3	0.4	0.5	0.6	0.7	0.8	0.9	1
Controls	0.00	0.00	0.51	9.18	19.90	21.43	27.04	16.84	4.59	0.51
Cases	0.00	1.20	3.61	10.64	19.88	23.69	23.49	11.85	4.82	0.80

Percent European ancestry

Figure 6.1   Histograms of European ancestry proportions in type 2 diabetes participants and controls from Mexico and Colombia, using 67 AIMs.

Note: Black bars, type 2 diabetes; grey bars, controls.

$p=4 \times 10^{-7}$) and Colombians (34 percent correlation, $p=2 \times 10^{-19}$), and therefore could potentially confound associations between genetic ancestry and type 2 diabetes. In the Mexicans, socioeconomic status was strongly predictive of case-control status ($p=2 \times 10^{-10}$), and when socioeconomic status and ancestry were analyzed together, the association between non-European ancestry and type 2 diabetes was significantly attenuated (OR 0.17 [0.04–0.71], $p=0.02$) (table 6.3). In the Colombians, the association between socioeconomic status and type 2 diabetes was weaker, but still strongly significant due to the larger sample size ($p=8 \times 10^{-7}$). Inclusion of both socioeconomic status and ancestry in the model abolished the significance of the ancestry-phenotype relationship (OR 0.64 [0.19–2.12], p=0.46) (table 6.3). For both populations, inclusion of socioeconomic status in the model reduced the effect size as well as the statistical significance of the association between ancestry and type 2 diabetes (table 6.3). However, in each case, the effect size and statistical significance of the association between socioeconomic status and type 2 diabetes was little changed by accounting for ancestry (table 6.3).

As an alternative way to investigate the association between non-European ancestry and type 2 diabetes while accounting for socioeconomic status, we conducted a stratified analysis, in which we analyzed each socioeconomic status stratum separately. We excluded the eight Colombian samples in socioeconomic stratum 6 from this analysis, so that five strata were analyzed for each population. In Mexicans and Colombians, the association between ancestry and type 2 diabetes did not reach statistical significance (nominal p=0.01, corrected p=0.05 after accounting for five statistical tests) for any stratum (figure 6.2a, b). These results are consistent with the greatly reduced association between ancestry and type 2 diabetes when accounting for socioeconomic status (table 6.3).

We also conducted a stratified analysis in which we considered each of five ancestry strata (0–20, 20–40, 40–60, 60–80, or 80–100 percent European ancestry) and analyzed associations between socioeconomic status and type 2 diabetes within each stratum (figure 6.2c, d). For Mexicans, we obtained nominal p values of 0.003, 0.12, 0.00001, 0.11, and 0.01, each with a negative coefficient for socioeconomic status. For Colombians, we excluded ancestry stratum 0–20 percent (which contained only six samples, each with type 2 diabetes) and obtained nominal

Table 6.3 Association of non-European ancestry with type 2 diabetes in Latinos is confounded by socioeconomic status

Country of origin	c_{SES} Logreg coefficient	Pseudo-r^2	p value	Logreg coefficients c_{SES}	c_{anc}	Pseudo-r^2 $c_{SES}+c_{anc}$	p values c_{SES}[a]	c_{anc}[b]
Mexico	−0.95	0.18	2×10^{-10}	−0.85	−1.79	0.20	9×10^{-8}	0.02
Colombia	−0.41	0.03	8×10^{-7}	−0.39	−0.45	0.03	1×10^{-5}	0.46

Notes: We report results of logistic regressions using socioeconomic status (c_{SES}) or socioeconomic status+ancestry ($c_{SES}+c_{anc}$) as predictors of disease outcome. $c_{subscript}$ refers to the logistic regression coefficients for each variable.

[a] With ancestry as covariate

[b] With socioeconomic status as covariate

p values of 0.004, 0.0006, 0.006, and 0.49, with a negative coefficient for socioeconomic status for each of the three significant p values. These results are consistent with the association between socioeconomic status and type 2 diabetes remaining strong after accounting for ancestry (table 6.3).

We finally considered whether BMI confounds the observed associations between either socioeconomic status or non-European ancestry and

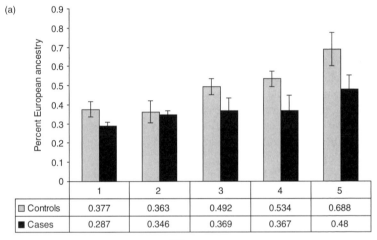

(a)

	1	2	3	4	5
☐ Controls	0.377	0.363	0.492	0.534	0.688
■ Cases	0.287	0.346	0.369	0.367	0.48

SES stratum

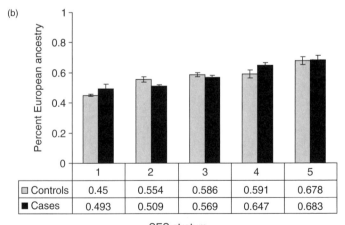

(b)

	1	2	3	4	5
☐ Controls	0.45	0.554	0.586	0.591	0.678
■ Cases	0.493	0.509	0.569	0.647	0.683

SES stratum

Figure 6.2   Continued

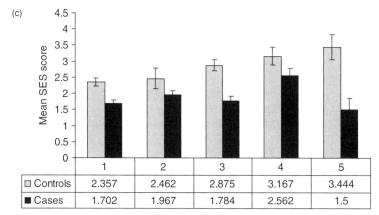

(c)

	1	2	3	4	5
☐ Controls	2.357	2.462	2.875	3.167	3.444
■ Cases	1.702	1.967	1.784	2.562	1.5

Ancestry stratum

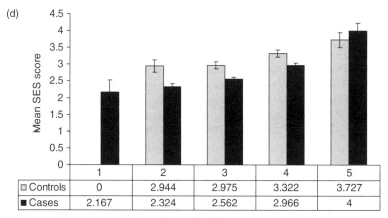

(d)

	1	2	3	4	5
☐ Controls	0	2.944	2.975	3.322	3.727
■ Cases	2.167	2.324	2.562	2.966	4

Ancestry stratum

Figure 6.2 Distribution of ancestry proportions across strata of socioeconomic status (SES) in (a) Mexicans and (b) Colombians. (c) Distribution of mean socioeconomic status scores across strata of increasing proportion of European ancestry in Mexicans and (d) in Colombians.

Note: Error bars denote standard errors of the mean. Black bars, type 2 diabetes; grey bars, controls

Table 6.4   Association of socioeconomic status and non-European ancestry with type 2 diabetes is not confounded by BMI

Country of origin	Logreg coefficients			Pseudo-r2	p values		
	c_{SES}	c_{anc}	c_{BMI}	$c_{SES}+c_{anc}+c_{BMI}$	c_{SES} [a]	c_{anc} [b]	c_{BMI} [c]
Mexico	−0.81	−1.72	0.08	0.22	6×10^{-7}	0.02	0.03
Colombia	−0.37	−0.46	0.06	0.05	3×10^{-5}	0.46	0.002

Note: We report results of logistic regressions using socioeconomic status (c_{SES}) + ancestry (c_{anc}) + BMI (c_{BMI}) as predictors of disease outcome. $c_{subscript}$ refers to the logistic regression coefficients for each variable.

[a]   With ancestry and BMI as covariates
[b]   With socioeconomic status and BMI as covariates
[c]   With socioeconomic status and ancestry as covariates

type 2 diabetes. We determined that BMI is negatively correlated with socioeconomic status in Mexicans (−20 percent correlation, $p=0.002$) and Colombians (−7 percent correlation, $p=0.05$), but that correlations between BMI and European ancestry proportion are not statistically significant in Mexicans (−12 percent correlation, $p=0.06$) or in Colombians (−1 percent correlation, $p=0.72$). The significant negative correlation between BMI and socioeconomic status implies that BMI could potentially confound our analyses. However, inclusion of BMI as a covariate in our analyses did not change the effect size or statistical significance of the associations of socioeconomic status and non-European ancestry with type 2 diabetes either in Mexicans or in Colombians, even though BMI itself was associated with type 2 diabetes in both populations (table 6.4).

Discussion

These data demonstrate a genetic association between type 2 diabetes and individual non-European ancestry proportions in Latinos, while also showing that this evidence of association is highly confounded by socioeconomic status. Combining our study with previous results, this pattern has now been observed in North American, Central American, and South American populations (Chakraborty et al. 1986; Parra et al. 2004; Martinez-Marignac et al. 2007), with similar trends also noted in African Americans (Reiner et al. 2005).

Our study design has several limitations. First, our strategies for estimating socioeconomic status are necessarily imprecise and differ between both locations. Second, because type 2 diabetic participants and controls were ascertained separately, some bias may have been introduced despite best efforts to match recruitment procedures. Third, our sample size may have been too small to assess the relative effects of highly correlated, potentially confounding variables. And fourth, given the diversity inherent in Latinos, our findings may not be generalizable to all Latinos or to other populations with admixed Native American ancestry. Nevertheless, we believe we have taken analytical measures to address potential confounders, and that the similar results obtained in two different populations strengthen our conclusions.

Our results show that, due to the correlation between socioeconomic status and Native American ancestry, it is difficult to disentangle the relationship between genetics and social factors in the contribution to disease risk. Low socioeconomic status can increase diabetes risk via a variety of mechanisms such as poor access to care, neglect of preventive strategies, a lower ability to exercise, or an unhealthy diet. The question of whether the increased susceptibility to type 2 diabetes in Native American populations is caused by genetic or social factors (or a combination of both) is difficult to resolve accurately as long as low socioeconomic status is more prevalent among persons with greater Native American ancestry; answering it appropriately would require that admixed case-control cohorts be carefully matched for socioeconomic status, a much larger sample be used, or twin studies discordant for socioeconomic status be undertaken. In some cases (e.g., the *ABCA1* R230C polymorphism), the strong association between a genetic variant and type 2 diabetes risk may be largely unaffected when adjusted for different confounders, including educational level as a surrogate of socioeconomic status (Villarreal-Molina et al. 2007; Villarreal-Molina et al. 2008). On the other hand, due to the strong association between Native American ancestry and socioeconomic status, use of the latter as a covariate in admixture studies might mask a true association signal in populations where low socioeconomic status is highly correlated to ancestral background.

The association between type 2 diabetes and Native American origin remained nominally significant after adjustment for socioeconomic status in the Mexicans, but not in the Colombians. This may be due to a weaker initial signal in the Colombians, differences in socioeconomic status ascertainment between the two populations, or a combination of both factors. Because the Mexican sample was enriched for early-onset cases ($n=81$), it is possible that genetic susceptibility to type 2 diabetes may have been stronger among the Mexicans. Another potential reason for the stronger *p* value in Mexicans is the very different distribution of Native American heritage in Mexicans versus Colombians (figure 6.1a, b). Mexicans have a wide range of ancestry proportions, so enrichment of the type 2 diabetes group by individuals with more Native American ancestry could cause a wider separation of ancestry proportions that is easier to measure. Conversely, Colombians have a narrower range, which may be due to more generations of mixing and homogenization (Bedoya et al. 2006). Just as in African Americans (Smith et al. 2004), this phenomenon limits the separation in ancestry proportion between cases and controls, even if a strong effect of ancestry causes oversampling of people who have inherited more of one ancestry. This may explain some of the discrepancies observed in the literature (Parra et al. 2004). Thus it is possible that Pima Indians, like the samples in this study from central Mexico, have a wide range of ancestry proportions, making the oversampling of individuals with more extreme values of one ancestry easily detectable, whereas the San Luis Valley samples, like our samples from Colombia, have a narrow spread of ancestries making the effect more subtle. A higher degree of admixture and homogenization in a society may also lead to decreased health disparities of a social nature. Alternatively, the lack of precision inherent in estimations of socioeconomic status (which often rely on information provided by participants who may have secondary motives for reporting a different stratum) may also have contributed to the observed differences. Our findings also illustrate the difficulties in making generalizations about Latino populations, which are often characterized by very diverse ancestral origins and environmental contexts that may affect disease, socioeconomic status, and their interactions.

These results also have implications for the prospects of gene mapping to find risk factors for type 2 diabetes in Latinos. Based on the epidemiological observation that type 2 diabetes is more prevalent in Latinos than in populations of European descent, it has been speculated that genetic risk factors for type 2 diabetes must be more common in Native Americans than in Europeans. Such variants could be located by admixture mapping, a technique that scans through the genome of affected individuals of mixed Native American, European, and African ancestry, searching for regions with unusually high proportions of one ancestry compared with the genome-wide average. The observation that increased type 2 diabetes prevalence in Latinos is at least partially explained by environmental factors decreases the likelihood that significant genetic risk factors can be easily found through this approach, although it does not rule out the possibility that the method can work. These results do not imply that genome-wide association scans cannot detect variants associated with either increased or decreased Native American ancestry. Indeed, when a variant is much more common in one population than another, it will be easier to achieve genome-wide statistical significance in a genome-wide scan performed in the former than it will be in the latter (McCarthy 2008; Unoki et al. 2008; Yasuda et al. 2008).

It is important to recognize that socioeconomic status will not be a confounder in type 2 diabetes admixture mapping in the same way as it is for the present analysis. In admixture disease mapping, each locus in the genome is separately tested for association, using the rest of the genome as a control to assess whether the locus stands out. If an association is observed at any locus, it must signify a real genetic connection of that particular locus to disease (socioeconomic status is not expected to be locus specific). However, because of the strong effect of socioeconomic status on type 2 diabetes risk in Latinos, rich information on socioeconomic status may be an important covariate that will increase statistical power in scans for genetic risk factors for type 2 diabetes. Tests for interactions between genes and environment in Latinos with type 2 diabetes may offer more power to detect risk factors for the disease than would be afforded if the modulatory effect of socioeconomic status on genetic risk were not taken into account.

Acknowledgments

We thank all the participants for their contribution of phenotype and DNA samples, and S. Jiménez-Ramírez for technical assistance. J. C. Florez is supported by a Massachusetts General Hospital Physician Scientist Development Award and a Doris Duke Charitable Foundation Clinical Scientist Development Award. A. L. Price was supported by a Ruth Kirschstein National Research Service Award from the National Institutes of Health (NIH). R. Saxena is supported by a Ruth Kirschstein National Research Service Award from the NIH. D. Reich is supported by a Burroughs Wellcome Career Development Award in the Biomedical Sciences. Support for this project was provided by discretionary funding from Harvard Medical School (to D. Reich), NIH grants NS043538 (to A. Ruiz-Linares) and DK073818 (to D. Reich), Colciencias grants 1115-04-16451 and 1115-04-012986 (to G. Bedoya and A. Villegas), and a Universidad de Antioquía grant (CODI/sostenibilidad 9889-E01321) to G. Bedoya and A. Villegas.

Note

First published in *Diabetologia*, 52, pp. 1528–1536, 2009

References

Bedoya, G. et al. (2006) Admixture dynamics in Hispanics: A shift in the nuclear genetic ancestry of a South American population isolate. *Proceedings of the National Academy of Sciences of the United States of America*, 103, pp. 7234–7239.

Bennett, P. H. et al. (1971) Diabetes mellitus in American (Pima) Indians. *Lancet*, 2(7716), pp. 125–128.

Carter, J. S. et al. (1996) Non-insulin-dependent diabetes mellitus in minorities in the United States. *Annals of Internal Medicine*, 125, pp. 221–232.

Chakraborty, R. et al. (1986) Relationship of prevalence of non-insulin-dependent diabetes mellitus to Amerindian admixture in the Mexican Americans of San Antonio, Texas. *Genetic Epidemiology*, 3, pp. 435–454.

Flegal, K. M. et al (1991) Prevalence of diabetes in Mexican Americans, Cubans, and Puerto Ricans from the Hispanic Health and Nutrition Examination Survey, 1982–1984. *Diabetes Care*, 14, pp. 628–638.

Gardner, L. I. et al. (1984) Prevalence of diabetes in Mexican Americans: Relationship to percent of gene pool derived from Native American sources. *Diabetes*, 33, pp. 86–92.

Haffner, S. M. et al. (1986) Hyperinsulinemia in a population at high risk for non-insulin-dependent diabetes mellitus. *New England Journal of Medicine*, 315, pp. 220–224.

Hamman, R. F. et al. (1989) Methods and prevalence of non-insulin-dependent diabetes mellitus in a biethnic Colorado population: The San Luis Valley Diabetes Study. *American Journal of Epidemiology*, 129, pp. 295–311.

Hamman, R. F. et al. (1978) Incidence of diabetes among the Pima Indians. *Advanced Metabolic Disorders*, 9, pp. 49–63.

Hanis C. L. et al. (1983) Diabetes among Mexican Americans in Starr County, Texas. *American Journal of Epidemiology*, 118, pp. 659–672.

Knowler, W. C. et al. (1990) Diabetes mellitus in the Pima Indians: Incidence, risk factors and pathogenesis. *Diabetes/ Metabolism Reviews*, 6, pp. 1–27.

Knowler, W. C. et al. (1988) Gm3;5,13,14 and type 2 diabetes mellitus: An association in American Indians with genetic admixture. *American Journal of Human Genetics*, 43, pp. 520–526.

Mao, X. et al. (2007) A genomewide admixture mapping panel for Hispanic/ Latino populations. *American Journal of Human Genetics*, 80, pp. 1171–1178.

Martinez-Marignac, V. L. et al. (2007) Admixture in Mexico City: Implications for admixture mapping of type 2 diabetes genetic risk factors. *Human Genetics*, 120, pp. 807–819.

McCarthy, M. I. (2008) Casting a wider net for diabetes susceptibility genes. *Nature Genetics*, 40, pp. 1039–1040.

Neel, J. V. (1962) Diabetes mellitus: A "thrifty" genotype rendered detrimental by "progress"? *American Journal of Human Genetics*, 14, pp. 353–362.

Parra, E. J. et al. (2004) Relation of type 2 diabetes to individual admixture and candidate gene polymorphisms in the Hispanic American population of San Luis Valley, Colorado. *Journal of Medical Genetics*, 41, p. e116.

Price, A. L. et al. (2007) A genomewide admixture map for Latino populations. *American Journal of Human Genetics,* 80, pp. 1024–1036.

Reich, D. E. et al. (2003) Quality and completeness of SNP databases. *Nature Genetics,* 33, pp. 457–458.

Reiner A. P. et al. (2005) Population structure, admixture, and aging-related phenotypes in African American adults: The Cardiovascular Health Study. *American Journal of Human Genetics,* 76, pp. 463–477.

Samet, J. M. et al. (1988) Diabetes, gallbladder disease, obesity, and hypertension among Hispanics in New Mexico. *American Journal of Epidemiology,* 128, pp. 1302–1311.

Silva Arciniega, M. R. and Brain Calderon, M. L. (2006) *Validez y confiabilidad del estudio socioeconómico.* Mexico City: Universidad Nacional Autónoma de México y Escuela Nacional de Trabajo Social.

Smith M. W. et al. (2004) A high-density admixture map for disease gene discovery in African Americans. *American Journal of Human Genetics,* 74, pp. 1001–1013.

Stern, M. P. et al. (1981) Cardiovascular risk factors in Mexican Americans in Laredo, Texas. II. Prevalence and control of hypertension. *American Journal of Epidemiology,* 113, pp. 556–562.

Tang, K. et al. (1999) Chip-based genotyping by mass spectrometry. *Proceedings of the National Academy of Sciences of the United States of America,* 96, pp. 10016–10020.

Tian, C. et al. (2007) A genomewide single-nucleotide-polymorphism panel for Mexican American admixture mapping. *American Journal of Human Genetics,* 80, pp. 1014–1023.

Unok, I. H. et al. (2008) SNPs in *KCNQ1* are associated with susceptibility to type 2 diabetes in East Asian and European populations. *Nature Genetics,* 40, pp. 1098–1102.

Villarreal-Molina, M. T. et al. (2008) Association of the ATP-binding cassette transporter A1 R230C variant with early-onset type 2 diabetes in a Mexican population. *Diabetes,* 57, pp. 509–513.

Villarreal-Molina, M. T. et al. (2007) The ATP-binding cassette transporter A1 R230C variant affects HDL cholesterol levels and BMI in the Mexican population: Association with obesity and obesity-related comorbidities. *Diabetes,* 56, pp. 1881–1887.

Williams, R. C. et al. (2000) Individual estimates of European genetic admixture associated with lower body-mass index, plasma glucose, and prevalence of type 2 diabetes in Pima Indians. *American Journal of Human Genetics,* 66, pp. 527–538.

Yasuda, K. et al. (2008) Variants in *KCNQ1* are associated with susceptibility to type 2 diabetes mellitus. *Nature Genetics,* 40, pp. 1092–1097.

7

Remembering or Forgetting Mendel: Sickle Cell Anemia and Racial Politics in Brazil

Peter Fry

They [sickle cell researchers] have appropriated sickling to support the particular interpretation of the relationship between nature and culture without which the notion of race would cease to make sense.

(Melbourne Tapper, 1999, 126)

Introduction

"We wish to affirm, and truly with considerable pride, our condition as a *multi-racial society* and that we have great satisfaction in being able to enjoy the privilege of having *distinct races* [*raças distintas*] and distinct cultural traditions also. In these days, such diversity makes for the wealth of a country" (my emphasis). These words were not pronounced by the president of South Africa, nor even by a multicultural zealot in Great Britain or the United States. Rather they were spoken on Brazil's Independence Day in 1995 by the recently elected president, sociologist Fernando Henrique Cardoso.

In a trice, Fernando Henrique had demolished that image of Brazil, which had emanated from all previous republican governments, namely Brazil as a society of mixture and hybridism. The old ideology of Brazil as a "racial democracy," widely attributed to the Brazilian anthropologist Gilberto Freyre, had been replaced by an anti-Freyrean discourse that had developed largely at the University of São Paulo where Fernando

Henrique wrote his doctoral dissertation on "race relations" in the south of Brazil (Cardoso, 1977) in close conjunction with black activists.

Later that year, Cardoso announced his new Human Rights Program (PNDH), which for the first time included reference to "racial inequality." To make the point as clear as possible, the new program was announced in Brasília on November 20, 1995, the three-hundredth anniversary of the death of Zumbi, who led the seventeenth-century maroon community of Palmares. Present were members of the black movement from all over Brazil who had traveled to the capital for the "Zumbi March against Racism and for Equality and Life."

The PNDH proposed strong measures to curb racism. State Secretaries of Public Security would be stimulated "to promote refresher courses and seminars on racial discrimination.... Adopt the principle of the criminalization of racism in the Penal and Code and Penal Process.... Disseminate the International Conventions, articles of the Federal Constitution and infra-constitutional legislation dealing with racism.... Support the production and publication of documents contributing to dissemination of anti-discriminatory legislation." But it went beyond these antidiscrimination measures to propose actions that were to be defined as positive discrimination or affirmative action. An Inter-Ministerial Working Group for the Recognition and Promotion of the Black Population (GTI) was created. The program proposed providing "support for private enterprises which undertake affirmative action [*discriminação positiva*]," developing "affirmative action to increase the access of blacks to professional courses, the university and areas of state of the art technology," and "formulating compensatory policies to promote the black community economically."[1]

Now, as I have argued elsewhere (Fry, 2005a), it is evident that affirmative action for "the black population" implies and requires the prior existence of such a population in logical and practical opposition to "the white population." And so it is not surprising that the Human Rights Program also "instructed" the Brazilian Institute for Geography and Statistics (IBGE), which is responsible for collecting official census data, to "adopt the criterion of considering mulattos, browns and blacks [*os mulatos, os pardos e os pretos*] as members of the black population [*integrantes do contingente da população negra*]." This instruction was never heeded, but it does show how the PNDH intended to bring the census categories in line with those of a Brazil imagined as a society of distinct races, by eliminating the intermediary category of "brown" (*pardo*), which covers such a multitude of phenotypical possibilities as to attract no less that 38.9 percent of the population.[2] In that way, Brazil would be measured indeed as a society of distinct races, through the remaining categories of white, black, yellow (this refers to Asians), and Amerindians.

Apart from the official census categories, two other taxonomies come into play in daily life for defining self and other in "racial" terms: a simple black/white one and another consisting of multiple descriptive terms. The taxonomy of multiple categories is dominant in the flow of most daily life. It is based fundamentally on the appearance of individuals, what Oracy Nogueira called the "mark" in contrast to the US system, which gives prominence to descent, which he denominated "origin" (Nogueira, 1959). It provides the language for most daily interactions and the way in which people define themselves and others from situation to situation. One survey conducted by the IBGE revealed over a hundred of them (Schwartzman, 1999). The bipolar one is predominant in the black movement, which has always had enormous difficulty in persuading all the shades of brown that they were *in fact* black and to adopt a black racial identity. The sociologists who defend such a taxonomy argue on statistical grounds that it makes more sense to aggregate the black and brown census categories than to keep them separate, arguing that the statistical gap between the whites and the nonwhites is of greater significance than any differences between browns and blacks. The black movement in Brazil traditionally used the previously pejorative word "Negro" to refer to themselves. Anthropologists had a strong tendency to use "Afro-Brasileiros," while more recently the activists have moved toward "afrodescendentes," which means literally "descended from Africans."

Racially based affirmative action, which is by definition bipolar, strengthens a Brazil that is imagined as made up of distinct racial groups.

Needless to say, state racialization and its attendant policies, which have grown in number and strength during the Lula government, have not gone unopposed, either in the writings of public intellectuals[3] or in the quotidian reaction of individuals who for one reason or another reject the new order.[4] Furthermore, there have been serious practical problems of deciding who is who, notably in the case of racial quotas in Brazilian universities (Maio and Santos, 2005). Notions of mixture as integral to national and individual identity continue to dominate the thought of many, including, it may be said, some supporters of affirmative action who argue that in the end Brazil's "racial cordiality" is too precious to be seriously challenged.[5] The debate, however, rages between those who defend formal government-sponsored racialization and those who oppose it. Those in favor can be divided into essentialists and constructionists. The former believe that Brazil is primordially divided into identity-sustaining races, which the new policies merely recognize. The latter believe that the dominant form of racial classification in Brazil discriminated all people of color, separating them from the whites.[6] All the other categories and notions of hybridism or mixture merely "mask" this "reality." They base their analysis on statistical and existential facts of inequality between people of lighter and darker hue and argue in favor of immediate if temporary pragmatic action. The opponents of state sponsored racialization are more radically constructivist. They argue that many ways of defining people in terms of color and race coexist in Brazil, but that enshrining two racial categories in the letter of the law will likely strengthen a bipolar racial system and will perpetuate rather than eliminate the insidious belief in races, which is, in the final analysis they claim, the prime cause of the reproduction of inequality.

Regardless of which side you may take on this political debate, the implementation of race-based affirmative action in Brazil has as one of its consequences the bringing into being of the premises upon which it is based as a self-fulfilling prophecy (Merton, 1948). Positing the existence of "distinct races," affirmative action policies bring into being discourses and practices that help maintain the illusion that this is the inescapable reality of Brazilian society. Nowhere is this clearer than in the discussion over health. While epidemiologists, sociologists, anthropologists, psychologists, and their like recognize the social dimension of sickness, and while religious people, in particular the followers of the myriad possession religions in Brazil, may interpret the onset of affliction as due in part to divine retribution or the workings of sorcerers, most of these and the population at large also associate illness with things natural. To enunciate "the health of the black population" is to posit (a) the existence of a discrete "black population" and (b) relate it to the natural world of health and sickness. It has the consequence of placing one of the rival taxonomies as part of the natural order of things, a fundamental demarche for the construction of its truth. (Durkheim and Mauss, 1969 [1903]).

The Health of the Black Population and Sickle Cell Anemia

In 1996, a workshop was convened by the subgroup of the GTI charged with looking into the health of the black population.[7] Composed of "scientists, civil society activists, doctors and technicians from the Ministry of Health," the workshop identified four groups:

- The first group is made up of genetically determined illnesses. They are illnesses, which share a hereditary, ancestral and ethnic cradle. In this group the most important illness is sickle cell anemia, because it is an illness which affects predominantly people of African descent [*afrodescendentes*]. Other illnesses or afflictions in this group are arterial hypertension, mellitus diabetes and a kind of hepatic enzyme deficiency, the Glucose-6-phosphate dehydrogenase, which affect other racial/ethnic groups, but are more serious or difficult to treat when they appear among blacks [*pretos*] and browns [*pardos*].

- The second block is made up of conditions, illnesses and afflictions derived from unfavorable socioeconomic and educational conditions, and also intense social pressure: alcoholism, drug addiction, malnutrition, high levels of infant morbidity, septic abortions, ferriprive anemia, STDs/AIDS, work related illnesses and mental problems.

- The third block is made up of illnesses whose development is worsened or whose treatment is made the more difficult on account of the bad conditions cited above: arterial hypertension, mellitus diabetes, coronaropathy, kidney insufficiency, cancers and myomas. This means that although these illnesses affect the population as a whole they are more serious in the black population, because of economic, social and cultural deprivation.

- The fourth block contains the set of physiological conditions which are affected by the above mentioned negative conditions, which contribute to their transformation into illnesses: growth, pregnancy, giving birth and aging. (Presidência da Rupública, 1998).

The group came to the conclusion that there was "no technical justification for the creation of a number of specific government health programs for the black population, as intended by some parts of the sector." The Sickle Cell Anemia Program (*Programa de Anemia Falciforme* [PAF]), however, in many ways became a flagship program for black activists and their supporters. But before looking more carefully at this program, and in order to put Brazilian black militant discourse on the ailment in perspective, I will spend just a little time looking at the relationship between sickle cell anemia and "race" in the United States through the medium of two excellent monographs, historian Keith Wailoo's *Dying in the City of the Blues: Sickle Cell Anemia and the Politics of Race and Health* (Wailoo, 2001) and anthropologist Melbourne Tapper's *In the Blood: Sickle Cell Anemia and the Politics of Race* (Tapper, 1999).

Sickle Cell Anemia in the United States

Since coming into being as a medical entity when it was diagnosed and named in 1910 in a student of Caribbean origin, sickle cell anemia has been associated with the "black race," despite its occurrence in people of Greek, Italian, Indian, and Latin American ancestry. Later theorists quite clearly associated the ailment with constitutional abnormality. "As they considered the appearance of sickled cells to be determined primarily by 'deeply rooted racial characteristics,' they transformed sickling from a mere *marker* of race to a *symptom* of race—more specifically racial inferiority" (Tapper, 1999, p. 39).

Tapper suggests that one

> would expect the work of these early sickling researchers to have lost currency after the genetic distinction between sickle cell trait and sickle cell anemia had been worked out in 1949. Neel's accomplishment should have changed the language in which sickling was being discussed, making the "old" racialist anthropology and eugenics irrelevant or outdated. In the context of the new genetics, on the manner of Neel and Linus Pauling, one would have expected such terms as race, the Negro (understood as a representative of a racial type), colored people, the subnormal individual, and status degenerativus to have become meaningless and to have been replaced by notions from molecular biology (Tapper, 1999, pp. 39–40).

But this would be to ignore the fact that paradigms die slowly if at all, especially when linked to massive social investment as Thomas Kuhn so brilliantly showed so many years ago (Kuhn, 1971). And this was indeed the case. "In the 1950s...eugenics and genetics did not exclude one another, but rather often co-existed or even fused so as to produce a new and powerful racialist anthropology that was informed—and authorised—by the language of molecular biology (especially the discourse on blood group genes)" (Tapper, 1999, p. 40). So strong was the relationship between sickling and race that it had become something of a closed system, producing protective secondary elaborations to account for observations that appeared to threaten it (Evans-Pritchard, 1965 [1937]). For example, patients defined as "white" could be easily reclassified as black. Sickling had become associated to such an extent with the black body that it became a defining characteristic of blackness. "My point," writes Tapper, "is that medical science and anthropology have used sickling to claim the racial distinctiveness of "blacks" and "whites" and to represent this distinctiveness as the product of a specific genetic structuration—as indisputable fact" (Tapper, 1999, p. 3).

The fusion of "race" and sickle cell anemia became so entrenched that in the 1970s, in the wake of the civil rights movement, the US government chose the control and prevention of sickle cell anemia as a measure to correct the history of segregation and discrimination against African

Americans. A striking feature of Keith Wailoo's history of sickle cell anemia in Memphis is the fact that it is taken for granted as a black person's ailment.

> To invoke the pain and suffering of the sickle cell patient was to dramatize the long-ignored social condition of black Americans and to give impetus to social activism. The malady propelled reform in health care and society. For white Americans, embracing the disease and acknowledging the legitimate pain of its sufferers could be an act of symbolic redemption (Wailoo, 2001, p. 7).

Tapper argues that in the end sickling became an emblem, the goal of full citizenship. Furthermore, as Tapper points out, an emphasis on individual responsibility grew and by extension a form of racial responsibility.[8] In an influential article, Robert B. Scott suggested that sickle cell anemia was the kind of social problem that could only be resolved were it to be combated by blacks themselves. Accordingly, he recommended genetic counseling for couples planning to marry. Tapper saw in this process the constitution of a new African American subject.

> The trait (and not the anemia) would ultimately provide a site for the surveillance of the reproductive behavior of African-Americans. More precisely, it became the vehicle for the constitution of the particular type of African-American subjects, who armed with knowledge—thanks to "the effectiveness of the educational and communication media of today," as Scott would have it—made informed decisions. These subjects, supposedly unlike previous generations of African-Americans, fulfilled the requirements for full citizenship and possessed knowledge of what they needed to do in order to serve their own interest and that of their community, and to act accordingly (Tapper, 1999, p. 106).

The 1972 National Sickle Cell Anemia Control Act, soon changed to the "National Sickle Cell Anemia *Prevention* Act," put responsibility firmly on the shoulders of the "black community," and Tapper wisely associates this demarche with the reaction to the Moynihan Report (Moynihan, 1965) that had described African American society as beset with family disintegration, criminality, and general anomie. Throughout the country, blacks organized themselves to take the test and to encourage others to so the same. Many states introduced compulsory tests for black citizens, and in some cases those with the trait lost work in some professions (air hostess for example). Others were barred from studying in military academies. Tapper argues convincingly that a significant consequence of the program was the celebration and even the constitution of a "black community" organized for the well-being of its members. "The dilemma," writes Tapper,

> facing the proponents of the sickling program was that...no well-defined, unified and well defined black community existed at the time of the hearings and the enactment of the National Sickle Cell Anemia Program Act; rather, it had to be produced. The proponents of the sickling initiative sought to assemble a positive image of that which was to be governed—the black community—as well as to stress government's good intentions, their presentation around such key terms as neglect (to be corrected), personal responsibility (presented as a universally available category to be appealed to), urgency (the insistence on immediacy being articulated as proof of the genuineness of government's concern for a group of marginalized citizens), and self-governance (as the crystallization of the benevolent nature of liberal democracy) determination. (Tapper, 1999, p. 123).

While Tapper's analysis brilliantly demonstrates the way in which the discourse around sickle cell anemia in effect brought into being its premise, namely a responsible black community, it fails to question the terms that preceded such action, namely, the racial categories of the United States (black, white, African American are just three of many), which appear as prediscursive concepts. This slip on the part of a confessed Foucauldian can only be interpreted in terms of the aura of naturalness that the bipolar racial taxonomy based on hypodescent has developed in the United States. That aura lurks almost imperceptibly behind almost all analysis of "race" not only in the United States but also, given its extraordinary global hegemony, not least Brazil, as I have argued elsewhere (Fry, 1995/96).

Turning to Brazil, then, where, as I have suggested at least two racial taxonomies vie for space in day-to-day life, each one with its own distinct political consequences, the discourse on sickle cell anemia that emerged in consort with black activism following the presidential declarations of 1995 may be understood not so much as a process of building responsibility in the black community, but rather as a process of strengthening and naturalizing its logical antecedent: the existence in Brazil of discrete black and white races and identities. I would argue that the dilemma facing the proponents of the Brazilian sickling program was that no well-defined black *population*, let alone *community* or *identity* existed at the time of it inception; rather, they had to be produced. As Gladys Mitchell has observed, "In the United States, racial consciousness was a precursor for politics, but in Brazil it appears that Brazilian policy changes such as affirmative action is a precursor for black racial consciousness" (Mitchell, 2004, p. 8).

Back to Brazil

As we have seen, sickle cell anemia was defined by the Working Group on the health of the black population as emanating from "an hereditary ancestral and ethnic cradle." The three adjectives "hereditary," "ancestral," and "ethnic" suggest that sickle cell anemia is a characteristic of a social group sharing a common ancestry and therefore the same genomic characteristics. The premise of a discrete ethnic *group* is strengthened throughout the Working Group's deliberations, which consistently refer to a *group* with distinct social characteristics in particular "economic, social and cultural deprivation." This latter concept, "cultural deprivation" is redolent of Moynihan's (Moynihan, 1965) understanding of black society in the United States of the 1950s and suggests that there is some form of cultural specificity to being black and poor in Brazil.

Even so, as I mentioned above, the only official program that emerged from the Working Group's deliberation was PAF.

In 2001, the Ministry of Health published its "Manual of illnesses important for ethnic reasons of the Brazilian population descended from Africans" (*Manual de Doenças mais Importantes, por Razões Étnicas, na População Brasileira Afro-Descendente*)" (Hamann and Tauil, 2001). The manual repeats much of the Working Group's findings, but also embarks upon a discussion of the supposed specificities of Brazilian blacks, in particular and above all because of the "intense miscegenation between Africans from various regions of Africa with the white population and, to a lesser degree ... with the indigenous population" (Hamann and Tauil, 2001, p. 9). After once again referring to the relative poverty of the black population, the manual concludes that "as far as illnesses with strong genetic determination are concerned, the Brazilian population of African descent [*afrodescendente*] may manifest them with particular characteristics, which means that it is not correct simply to transpose research results on these illnesses conducted in other countries" (Hamann and Tauil, 2001, p. 9). I am not sure whether the term "*afrodescendente*" was chosen for its possible genetic connotations or simply because it has recently become more politically correct than the terms "blacks" or "Afro-Brazilians." Probably the latter, for to state that there is a higher incidence of the sickling gene among descendants of Africans is not the same as saying the same thing for people who call themselves brown or black when visited by the Brazilian census takers!

Interestingly enough the supposedly greater "miscegenation" in Brazil is invoked not to define a possible specificity of the population as a whole, but of the "population descended from Africans." But such a possible national specificity would have no reason to exist in a document whose raison d'être is to define those illnesses "important for ethnic reasons of the Brazilian population descended from Africans." With no rules of "racial" endogamy and high rates of intermarriage and/or productive sex between people from Africa with others from Europe and from local Amerindian populations would signify that the sickling gene has permeated large swathes of the Brazilian population as is attested by all recent genetic research.

So far, it would seem then that as in the United States, the Mendelian logic of sickle cell anemia had been discretely neglected, allowing for the establishment of a simple and linear relationship between the illness and an "ethnic group." But the manual also contains a long article by Marco A. Zago, professor of clinical medicine of the Faculty of Medicine of Ribeirão Preto of the University of São Paulo (Zago, 2001). In this article, Zago recognizes that there is greater incidence of the sickling trait where there are more people who declare themselves black or brown but makes no reference to "race" or "ethnic groups." By recognizing the Mendelian logic of the sickness's transmission he proposes universalistic measures for its control. In fact he appeals more to problems associated with poverty than of "race." "To be efficient, strategies for controlling sickle cell illnesses should be associated with improvements in hygiene, public health and education in the most poverty stricken areas. (Zago, 2001, p. 30). He ends with a description of the PAF: "A set of activities designed to promote knowledge of the illness and its prevention, ease of access to diagnosis and treatment, as well as an educational campaign for the population at large and health professionals" (Zago, 2001, p. 30). These would include increasing the scope of diagnosis and treatment, capacity building for health professionals, and a search for alliances with "institutions and NGOs that operate in the area of sickle cell anemia." As we shall see, the majority of NGOs that joined the PAF were organizations of black activists, most of them women, who played a most important role in addressing the issue of sickle cell anemia as a problem of public health.

Sickle Cell Anemia and Black Activism

As we have seen, the origins of a concern with health and the black population have their ritual origins in a highly charged political event organized by black activists, the Zumbi March against Racism and for Equality and Life in 1995. It was also born at a point of inflection in government policy with regards to "race," when Brazil ceased to be seen by government as a land of mixture to become a society of "distinct races."

Although the PAF as described in the "Manual of illnesses important for the Brazilian population descended from Africans for ethnic reasons" contains no reference to black activism, it has been intimately linked to black activists in governmental institutions (most states and municipalities have councils for the advancement of the black community) and nongovernmental institutions. As the states of the Brazilian Federation implement their programs for sickle cell anemia, so information on sickle cell anemia as produced by these organizations reaches the far corners of Brazilian society.

In the state of Santa Catarina for example, the law that brought the PAF into being (State Law No. 12.131, March 12, 2002) guarantees "the participation of technicians and representatives of the black movement in the group which will be formed to implement the Program." In the state of Pará in the north of Brazil, a similar law (State Law No. 6.457, March 4, 2002) proposes "partnerships with the black movement." In the city of São Paulo, the wealthiest city of the federation, the Special Coordination of the Black (*Coordenadoria Especial do Negro—CONE*), which is responsible for "encouraging and developing policies to integrate the black in society, overcoming racial inequalities," developed a program called "Black City" (*Negra Cidade*) to combat racism and to guarantee racial diversity (*Combate ao Racismo e de Garantia da Diversidade Étnica*). One *Negro Cidade* program is designed to "reduce infant mortality caused by sickle cell anemia." (http://www.portalafro.com.br/entidades/cone/internet/cone .htm, accessed December 9, 2004). In Rio de Janeiro, the Nazareth Cerqueira Reference Centre against Racism and Anti-Semitism, which is part of the State Secretariat for Justice and Citizens' Rights, and whose principal function is to receive and process cases of racism, has regularly promoted a series of encounters to disseminate "information on the illness and improve the quality of life of sufferers from sickle cell anemia through integrated actions of the State and civil society." (http://www.sbhh.com.br /menu/noticias.asp?newsID=1293, accessed December 11, 2004).

Among the black NGOs concerned with sickle cell anemia, women activists have been predominant.[9] Psychologist Edna Roland, president of Fala Preta! Organização de Mulheres Negras (Speak out Black Woman! Organization of Black Women), was elected rapporteur général of the World Conference against Racism, Racial Discrimination, Xenophobia and Related Intolerance held in Durban in 2001. She has contributed to the discussion on the relations between "race" and women's reproductive health. As intellectual and as activist, she argues for the importance of hearing black women's voices who, through their organizations can demand from the state "the necessary conditions to be able to exercise their sexuality and reproductive rights, controlling their own fecundity" (Roland, 2001).

She founded Fala Preta!, whose folder "Anime-se Informe-se Anemia Falciforme" (Cheer up Get Informed Sickle Cell Anemia) was produced in cooperation with a major social science research institution in São Paulo, Centro Brasileiro de Análise e Planejamento (CEBRAP), and with the program on the "Reproductive Health of the Black Woman."

Jurema Batista, who founded the NGO "Crioula," is coauthor of a collection of essays on the health of black women (Werneck et al., 2000). Physician Fátima Oliveira argues that the field of health is an important arena for the fight against racism. She draws attention to discrimination against black patients and talks of the lack of "ethical and legal rigour" in governmental agencies and in laboratories where black patients are submitted to high-risk experiments.

A fourth prominent woman is Berenice Kikuchi. A qualified nurse with a master's degree in education, she is the founding president of the Sickle Cell Anemia Association of the state of São Paulo. Having contributed to writings on the illness in Brazil (Kikuchi, 1999, 2007), she also contributed actively in formulating the PAF.

In practice, then, governmental initiatives related to sickle cell anemia are, as they were in the United States, closely linked to black activists and their organizations, thus reinforcing the representation of the illness as somehow racial.

The Media

Débora Diniz and Cristiano Guedes have analyzed articles on sickle cell anemia in two Brazilian newspapers between 1998 and 2002, the *Folha de S. Paulo* (25 articles) and *A Tarde* from Salvador in the northeastern state of Bahia (41 articles). They were able to detect a strong association between the illness and the black population. "In most cases, the *A Tarde* newspaper described sickle cell anemia "primarily as a public health issue of interest to the black population, rather than as a genetic illness.... If it were possible to identify the principal theme which determined reporting on sickle cell anemia we would risk saying that it was the Brazilian black population" (Diniz and Guedes 2004, p. 1060). Concerned to discover the reasons for such an "equivocal" stance, they raised the hypothesis that the association of sickle cell anemia with the black population and therefore of differences between distinct identities has strong media impact:

> There must be very few biomedical specialists consulted by the newspapers who do not know that sickle cell anemia is not confined to blacks [*negros*] and browns [*pardos*], which make the analysis of the reasons for this equivocal stance even the more interesting. A possible explanation is that the

association between sickle cell anemia and blacks is a theme with strong media impact, strengthening the sociological expectation that genetics provides information on the origins of the individual. If this hypothesis is right, to confine sickle cell anemia to blacks and browns also means talking of their origins and differences (Diniz and Guedes, 2004, p. 1060).

This hypothesis becomes more plausible after looking at an important folder called "Sickle Cell Anemia—let us travel through this history…" (*Anemia Falciforme—viajemos por essa história…*) produced under the supervision of Berenice Kikuchi. It was published by the Sickle Cell Anemia Association of São Paulo, by the Special Coordination of the blacks in the municipality of São Paulo (*Coordenadoria Especial do Negro do Município de São Paulo*) and the Unified Black Movement (*Movimento Negro Unificado—MNU*), receiving support from other black organizations, such as the Hip Hop Movement, U Negro, the Strength and Race Group (*Grupo Força e Raça*), the Cultural Association Ruth de Souza (*Associação Cultural Ruth de Souza*), and the Collective of the Humanitarian league of Men of Color (*Coletivo da Liga Humanitária dos Homens de Cor*).

The cover of the folder shows a black nuclear family: father, mother, with their son who is playing football.[10] To their side is another character, also black, wearing a white coat with a stethoscope round his neck. Thus, right from the start the reader is introduced to an illness associated with black sufferers and healers.

The pamphlet in the form of a strip cartoon tells how Thiaguinho's parents learn that their son has sickle cell anemia, what this illness is, and how it can be treated. The story begins in Africa, a "rich and mysterious continent," "full of history in the cities" written over a pyramid and sphinx, "the simple life of the villages" over a bucolic village scene, "her wealth" over gold bars, and "her culture" over a mosque. This page ends with a black patch with the words: "And thus were the lives of the great grandparents of our characters."

Page two begins with a man on a floor mat, with the words: "The illness was known in various African nations, and knowledge of medicine was put to use to alleviate suffering." In the next picture a Portuguese caravel approaches the shoreline with the words: "One day, great ships began to arrive on the continent. They were foreigners in search of the wealth of the blacks (gold, spices, land) and also laborers." "And so the Africans were brought to Brazil from the days of the Discovery more than 400 years ago." In the next picture, we see a young man in the sugarcane fields. The last one depicts a political rally with placards reading "Zumbi," "Liberty," and "End slavery." "Here, they worked hard during the periods of sugar cane plantations, gold and coffee… and fought hard to regain their liberty...and to this very day they continue fighting for their citizenship and their liberty."

The next page begins with the doctor saying, "As descendents of those people it is important to know that the illness began there [in Africa]. But other peoples also suffer from the illness." This formulation is interesting for although the illness is not attributed exclusively to black people, others who may become ill are referred to as "peoples," as if sickle cell anemia were an illness always associated with ethnicity. Thiaguinho's parents reply: "Now we understand. That is to say that this illness has been passing from father to son for thousands of years up to our own days." The doctor replies: "Exactly. Now I am going to explain to you in detail how this occurs." The next page starts with a rather gruesome depiction of the sickle cells "blocking the circulation of the blood," and over a picture of a sad father is written: "And that is one of the reasons Thiaguinho suffers from pains, infection and anemia." The mother then asks how Thiaguinho can be sick while both she and her husband are healthy. The doctor then explains that those who receive the gene from only one parent remain healthy but carry the gene, while those who receive the gene from both parents develop the illness, which is the case of their son.

The next page explains how to test for the illness, affirms that it is not contagious, and that although it has no cure, Thiaguinho will be able to live a normal life "with medical assistance a few simple health precautions." Finally Thiaguinho appears with his high school diploma with the words "nothing can stop him becoming a great man." The back cover reproduces the Mendelian probability diagrams where the sickle cells gene is colored brown and the non-sickle cell gene, white. Whether through coincidence or not, we cannot tell.

The folder as a whole, with only black characters, with the exaltation of Africa and with the notion that sickle cell anemia is of "peoples" transmits a strong and almost exclusive relationship between sickle cell anemia, the black body, and a black identity. It recalls Wailoo's assertion for the United States, that the disease "became much more than a rallying point for polite local civic groups. It was remade into a national—and even an international—symbol of pain, social suffering, social inequality, and even the forceful uprooting of African Americans and the black diaspora." (Wailoo, 2001, p. 22)

Back to Comparison

The greatest similarity between the two situations, the United States in the 1970s and contemporary Brazil, is, of course, that both share an emphasis on the African origins of sickling and involve the enthusiasm of black activists and their allies.

The differences are maybe more significant. First, it is important to observe the preeminence of black women in the Brazilian case. Second, while Tapper mentions the presence of a strong black criticism of the US program—that it could lead to increased stigmatization of blacks, for example—I have no notice of any criticism from any black leadership. On the contrary, the folder I have described has the surprising effect of transforming a gruesome ailment into an emblem of racial pride and the black activists understand the PAF as a "conquest of the black movement."

But perhaps the greatest difference is that whereas in the United States, the discourse around sickle cell anemia was characterized by an emphasis on the bringing into being of a "responsible black community," in Brazil, the center of the discourse was the emphasis on the intimate relationship between the ailment and "the black population." In the United States, the categories "white" and "black" are taken as natural in accordance with hypodescent through the one-drop rule. Of course there can be false whites, who are "truly" blacks who "pass," but these serve only to shore up the basic model. There is little ambiguity in the system, and it is strong enough to have survived recent attempts to bring in new categories such as "biracial," which in itself betrays a commitment to two basic races (Daniel, 2006).

The Brazilian program seems to have another consequence: the naturalization of the logical antecedent of a "responsible black community," namely the "black race" itself. I suggest that the apparent total support of black activists for PAF signifies that sickle cell anemia represents much more than a disease to be eradicated. It is also a "natural symbol" of the "black race," and through logical and political opposition, the white "race" in Brazil.

As I suggested at the beginning of this paper, anti-Freyrean sociologists and black activists (these are not of course discrete categories) have aimed to have Brazilians imagine themselves no longer as a cordial genetic and cultural mixture of peoples, but as two tensely related races. But even so, and in spite of affirmative action policies that depend on a clear distinction between blacks and whites, the black movement continues to run up against the difficulty of persuading Brazilians to see themselves as either black or white. Maybe this allows us to understand why black activists and their sympathizers took the lead in the sickle cell program. They, more than anyone else, believe that Brazil is *in reality* a society of blacks and whites and must feel the frustration of not having been able to convince enough browns and mulattos to join common cause in a mass political movement. Their emotional and political commitment to changing the existing order must surely contribute to their preference for a racial/ethnic understanding of sickle cell anemia.

But this is only part of the story. "Brazil is not a country for amateurs," as singer and composer Tom Jobim is reputed to have said. For, even though the black activist NGOs have dominated the civil society participation in the PAF with their talk of races and ethnic groups, Mendel has not been ignored in the practice of diagnosis and treatment. As mentioned earlier, where the PAF has been introduced, all babies, regardless of their color, are tested for the sickling gene.

For some years now, since at least 2000 with the publication of "Retrato Molecular do Brasil" in 2000 (Pena et al., 2000), Brazilian geneticists have given a new legitimacy to the old Brazilian self-image as a society that developed out of the mixture of Europeans, Africans, and Amerindians. More recently, J. Alves-Silva, Sérgio Pena, and others have insisted on the fact that phenotype conveys little information about genotype (Parra et al., 2003; Pena, 2008). At least 40 percent of the genetic markers found in Brazilians who see themselves as white are of African origin, for example (Alves-Silva et al., 2000; Pena, 2002). Pena goes as far as to suggest that concepts of "race" should be entirely banned from the practice of medicine (Pena, 2005).

Even if sickle cell anemia appears with greater frequency among populations with more African phenotypes, it also affects people who in no way appear to have any African ancestry. The inverse is also true. Research in 24 black rural communities descended from runaway slaves in five regions of Brazil revealed that the frequency of the sickling gene varied from zero in the community of Itamoari in the northern state of Pará, and Paredão in the southern state of Rio Grande do Sul, to 13 percent in Riacho de Sacutiaba in the northeastern state of Bahia (Pedrosa et al., 2004). The researchers concluded that "considering all the descendants of runaway slave that we studied, the average frequency of the [sickling] allele was 3.7%, well below the 8.7% that has been found in those African countries where the majority of slaves were captured" (Pedrosa et al., 2004, p. 85). The authors of the research conclude that "it is not possible, in the case of sickle cell anemia, to establish a single health policy for all these populations. Each one should receive individualized attention in accordance with its specific characteristics" (Pedrosa et al., 2004, p. 85).

In spite of the fact that the black activists involved in the implementation of the PAF have insisted on a relationship between sickle cell anemia and the black population, the opinion of the geneticists has predominated in diagnosis and treatment.

In contrast again with the United States, where in some states only people defined as black were tested for the gene, the Brazil program aims to universalize the test in newborn babies, regardless of the color of their parents. In the state of Santa Catarina, for example, the law states: "The exam for diagnosing hemoglobinpathies will be made available to all new Born babies, and will take place in all maternity wards and similar hospitals in the State and throughout the National Unified Health System."

So, even though the PAF has consistently emphasized a firm relationship between sickle cell anemia and the black population, medical practice has followed the discourse of the geneticists so that, for the good of public health in Brazil, this ailment is understood to threaten any citizen, regardless of appearance or identity.

But more than that, geneticists such as Sergio Pena have taken a prominent position that is critical of the increased governmental racialization of Brazilian society. In his chapter in this book, Pena and his philosopher colleague Telma Birchal argue that because the genome of each Brazilian is highly variable "it does not make sense to talk about Afro-descendants or Euro-descendants... the only way of dealing scientifically with the genetic variability of Brazilians is individually, as singular and unique human beings in their mosaic genomes and in their life." Going one step further, the authors then go on to affirm that this scientific discovery can and should illuminate social thought and political action: "When implementing well-intentioned programs of affirmative action to push the necessary social changes forward, the government must be careful not to stir up artificial and arbitrary tensions and divisions among the people of Brazil, a country where, essentially, we are all equally different."

Contemporary genetics are thus brought in to lend a natural aura to what Louis Dumont termed the modern individual, very much a social and historical construct of the West (Dumont, 1972, 1986). Those who conveniently ignore Mendel are able to imagine Brazil as a society of natural and distinct races. Those who remember him and develop their research into the genomics of the Brazilian population are able to bring their genetic findings to support a Brazil that is imagined as a republican society of socially, but now also *naturally* discrete and autonomous individuals.

Notes

1. A presidential decree promulgated on November 20, 1995, created a Ministerial Working Group charged with "developing policies to promote the Black population" (GTIVPN, 1995).
2. The 2000 census figures put whites at 53.4 percent, blacks at 6.1 percent, yellows at 0.5 percent, and Amerindians at 0.4 percent.

3. There is quite an extensive bibliography, but the most complete source of such writings is Fry et al. (2007).

4. In an interesting article, Baran (2007) describes the difficulties of a school teacher in Bahia who attempts to convince students that they should see themselves as black and white.

5. "Inter-racial cordiality—both as reality and as myth—is still celebrated as something too precious to be threatened by the measures which haven't all that much impact anyway." (Sansone, 2006).

6. For a more detailed discussion of these two positions, see Grin (2010) and Fry (2009).

7. Much information and the basic argument are to be found in detail in Maio and Monteiro (2005).

8. "The management of sickling involved a sort of intensive care of the self and by extension of the race." (Tapper, 1999, p. 97).

9. I am very grateful to Consolação Lucinda for her help on this section. She drew my attention to the importance of these women and brought to my attention the folder called "Anemia Falciforme—viajemos por essa história."

10. The images of this folder are reproduced in Fry (2005b, pp. 360–361) (http://www.scielo.br/pdf/hcsm/v12n2/06.pdf accessed on April 10, 2011.

References

Alves-Silva, J. M. et al. (2000) The ancestry of Brazilian mtDNA lineages. *American Journal of Human Genetics*, 67, pp. 444–461.

Baran, M. D. (2007) "Girl, you are not morena. We are negras!": Questioning the concept of "race" in southern Bahia, Brazil. *Ethos*, 35(3), pp. 383–409.

Cardoso, F. H. (1977) *Capitalismo e escravidão no Brasil meridional: O negro na sociedade escravocrata do Rio Grande do Sul*. Rio de Janeiro: Paz e Terra.

Daniel, G. R. (2006) *Race and Multiraciality in Brazil and the United States: Converging Paths?* University Park, Pennsylvania: The Pennsylvania State University Press.

Diniz, D. and Guedes, C. (2004) A informação genética na mídia impressa: A anemia falciforme em questão. *SérieAnis*, 35, pp. 1–7.

Dumont, L. (1972) *Homo Hierarchicus: The Caste System and Its Implications*. London: Paladin.

———. (1986) *Essays on Individualism: Modern Ideology in Anthropological Perspective*. Chicago and London: The University of Chicago Press.

Durkheim, E. and Mauss, M. (1969 [1903]) *Primitive Classification*. London: Cohen & West.

Evans-Pritchard, E. E. (1965 [1937]) *Witchcraft, Oracles and Magic Among the Azande*. Oxford: Claredon Press.

Fry, P. (1995/96) O que a cinderela negra tem a dizer sobre a "política racial" no Brasil. *Revista USP* (São Paulo), pp. 122–135.

———. (2005a) *A persistência da raça: Ensaios antropológicos sobre o Brasil e a África austral.* Rio de Janeiro: Civilização Brasileira.

———. (2005b) O significado da anemia falciforme no contexto da "política racial" do governo brasileiro 1995–2004. *Historia, Ciencia e Saude-Manguinhos,* 12(2), pp. 347–370.

———. (2009) Anthropologists and the dispute over the meaning of race in Brazil. *Lusotopie, 16* (2), pp. 185–204.

Fry, P., et al. (eds.) (2007) *Divisões perigosas: Políticas raciais no Brasil contemporâneo.* Rio de Janeiro: Civilização Brasileira.

Grin, M. (2010) *"Raça": Debate público no Brasil.* Rio de Janeiro: Mauad and Faperj.

GTIVPN (Grupo de Trabalho Interministerial para Valorização da População Negra). (1995) Construindo a democracia racial. Available from http://www .biblioteca.presidencia.gov.br/area-presidencia/pasta.2008-10-08.1857594057/pasta.2008-10-08.9262201718/pasta.2008-12-16.0710539708/ pasta.2009-08-15.1277050055/Construindo%20a%20democracia%20racial .pdf.

Hamann, E. M. and Tauil, P. L. (2001) *Manual de doenças mais importantes, por razões étnicas, na população brasileira afro-descendente.* Brasília: Ministério da Saúde.

Kikuchi, B. A. (1999) *Anemia falciforme manual para agentes de saúde e educação.* Belo Horizonte: Editora Health.

———. (2007) Assistência de enfermagem na doença falciforme nos serviços de atenção básica. *Revista Brasileira de Hematologia e Hemoterapia,* 29(3), pp. 331–338.

Kuhn, T. (1971) *The Structure of Scientific Revolutions.* Chicago: The University of Chicago Press.

Maio, M. C. and Monteiro, S. (2005) Tempos de racialização: O caso da "saúde da população negra" no Brasil. *História, Ciências, Saúde-Manguinhos* 12 (2), pp. 419–446.

Maio, M. C. and Santos, R. V. (2005) Políticas de cotas raciais, os olhos da sociedade e os usos da antropologia: O caso do vestibular da Universidade de Brasília (UnB). *Horizontes Antropológicos, 23,* pp. 181–214.

Merton, R. K. (1948) The self-fulfilling prophecy. *Antioch Review, 8,* pp. 193–210.

Mitchell, G. (2004) Whitening, color identification, and support for affirmative action in Brazil. *Political Science.* Chicago: University of Chicago

Moynihan, D. P. (1965) *The Negro Family: The Case for National Action.* Washington, D.C.: United States Department of Labor.

Nogueira, O. (1959) Skin color and social class. In: Rubin, V. (ed.) *Plantation Systems of the New World.* Washington, D.C.: Research Institute for the Study of Man and the Pan American Union, pp. 164–179.

Parra, F. C. et al. (2003) Color and genomic ancestry in Brazilians. *Proceedings of the National Academy of Sciences of the United States of America, 100,* pp. 177–182.

Pedrosa, M. A. et al. (2004) Anemia falciforme em antigos quilombos. *Ciência Hoje,* 211, pp. 84–85.

Pena, S. D. J. (2005) Razões para banir o conceito de raça da medicina brasileira. *Historia, Ciencia e Saude-Manguinhos,* 12(2), pp. 321–346.

———. (2008) *Humanidade sem Raças?* São Paulo: Publifolha.

———. (ed.) (2002) *Homo brasilis: Aspectos genéticos, linguisticos, históricos e socio-antropológicos da formação do povo brasileiro.* Ribeirão Preto: FUNPEC.

Pena, S. D. J. et al. (2000) Retrato Molecular do Brasil. *Ciência Hoje* 159, pp. 16–25.

Presidência da República (1998) *Construindo a democracia racial.* Available from: https://www.planalto.gov.br/publi_04/COLECAO/RACIAL.HTM

Roland, E. (2001) Saúde reprodutiva da população negra no Brasil: Um campo em construção. *Perspectivas em Saúde e Direitos Reprodutivos,* 2(4), pp. 17–23.

Sansone, L. (2006) O bebê e a água do banho—a ação afirmativa continua importante não obstante os erros da UnB! In: Steil, C. (ed.) *Cotas raciais na universidade um debate.* Porto Alegre: Editora UFRGS, pp. 93–97.

Schwartzman, S. (1999) Fora de foco: Diversidade e identidades étnicas no Brasil. *Novos Estudos Cebrap,* 55, pp. 83–96.

Tapper, M. (1999) *In the blood: Sickle cell anemia and the politics of race.* Philadelphia: University of Pennsylvania Press.

Wailoo, K. (2001) *Dying in the City of the Blues Sickle Cell Anemia and the Politics of Race and Health.* Durham and London: University of North Carolina.

Werneck, J. et al. (2000) *O livro da saúde das mulheres negras: Nossos passos vêm de Longe.* Rio de Janeiro: Pallas.

Zago, M. (2001) Anemia falciforme e doenças falciformes. In: Hamann, E. M. and Tauil, P. L. (eds.) *Manual de doenças mais importantes, por razões étnicas, na população brasileira afro-descendente.* Brasília: Ministério da Saúde, pp. 13–35.

Part III

Genetics, History, Nationhood, and Identity

8

Gene Admixture Analysis through Genetic Markers and Genealogical Data in a Sample from the Buenos Aires Metropolitan Area

Francisco R. Carnese, Sergio A. Avena,
Maria L. Parolin, Maria B. Postillone, and
Cristina B. Dejean

Introduction

Several cosmopolitan populations of Argentina have been studied to estimate the European, Amerindian, and African participation in their gene pool. In the city of Buenos Aires, these studies were first made using only blood group markers such as ABO and D factor of Rh system (a revision of these studies was performed by Avena 2003). In addition, Sergio Avena et al. (1999, 216) carried out a comparative analysis between the current data and those available at a historical level using the frequencies of ABO and Rh system, and they were able to find significant differences between the sample of the Hospital de Clínicas of the city of Buenos Aires (register of donors from 1994 to 1995) and the data registered in the city in 1938 and 1949 (see Etcheverry 1949, 167; Avena 2003, 48). In that study, they observed an increase in ABO*O and RH*D alleles due to the important contribution of Amerindian people to the gene pool of the Buenos Aires Metropolitan Area (BAMA) population. Later, we determined ten blood group systems and gm/km immunoglobulins in two samples of the BAMA, and we obtained similar gene admixture values of 14–16 percent and 3.3–4.0 percent with Amerindians and Africans, respectively (Avena et al. 2001, 86; Avena et al. 2006, 115). Furthermore, when the sample of the BAMA was divided into three areas (city of Buenos Aires, first outer ring, and second outer ring) according to the place of present residence of the donors, the percentage of indigenous participation increased from the city of Buenos Aires to the first and second outer-ring suburbs, with percentages of 5, 11, and 33 percent, respectively (Avena 2003, 99); whereas the African components did not present significant variation between these areas, given that the range of variation was from 3.5 percent to 5 percent.

At another level of analysis, the same sample was studied by Fejerman et al. (2005, 168), who determined 12 molecular autosomal markers and obtained 2.2 percent of gene admixture with Africans. In this study, they did not determine the mixture with Amerindian people.

As we can see, in the city of Buenos Aires, all these studies have been performed through the determination of protein and molecular markers of biparental transmission. The results obtained have demonstrated that the population's genetic composition in this area is a product of contributions coming from several geographic groups. However, we did not study the distribution and frequencies of uniparentally inherited genetic markers such as mitochondrial DNA (mtDNA) and Single Nucleotide Polymorphisms (SNPs) located in the nonrecombinant Y chromosome. For this reason, in this study we now analyze the degree of participation of paternal and maternal European, Amerindian, and African lineages in the same sample of the BAMA population mentioned above, which had also been examined at a biparental level. Likewise, we study the regions that have contributed with those lineages and in what proportion they have done so. In addition, we analyze the influence that the donors' grandparents' marriages had on the distribution of those lineages and evaluate if these biological unions were or not preferential according to the couples' geographic origins. This information will allow us to estimate the proportion of admixture among them and to notice if this mixing process took place at that moment or if it occurred before, in the places of origin of the donors' grandparents.

Population Studied

Eight censuses carried out in the city of Buenos Aires from the end of the eighteenth century and during the nineteenth century (1778, 1806, 1810, 1822, 1827, 1836, 1838, and 1887) were analyzed (table 8.1). It was observed that between 1778 and 1838 the population increased almost threefold from 24,363 to 62,957 inhabitants.

Table 8.1   Results of eight censuses of Buenos Aires city, 1778–1887

Year	White people	Indians or mestizos	Afroargentines	Undetermined	Total	% of Afroargentines (*)	% of Indians or mestizos
1778	16,023 (66%)	1,104	7,235	—	24,363	29.7	4.5
1806**	15,078 (59%)	347	6,650	3,329	25,404	30.1	1.6
1810	22,793 (70%)	150	9,615	—	32,558	29.5	0.5
1822	40,616 (73%)	1,115	13,685	—	55,416	24.7	2
1827**	34,067 (80%)	152	8,321	—	42,540	19.5	0.4
1836	42,445 (67%)	—	14,906	6,684	63,035	26	—
1838**	42,312 (67%)	—	14,928	5,717	62,957	26.1	—
1887	425,370 (98%)	—	8,005	—	433,375	1.8	—

Notes/Source: (*) Seventh column was calculated without the data of the undetermined ones.

** Incomplete results / (Andrews 1989).

As regards the total population, the "white" people presented some fluctuation with a range of variation from 59 percent in 1806 to 80 percent in 1827. On the other hand, the black/mulattos decreased from 30 percent in 1806 to 20 percent in 1827. The Indian and mestizo population was small, and they were mentioned only in the first five censuses. They had a variation from 4.5 percent in 1778 to 0.4 percent in 1827, which may indicate a declining trend in the Indian population. However, all these values have to be considered with caution, given that in the censuses of 1806, 1836, and 1838, there were a total of 3,329, 6,684, and 5,717 individuals who were not included in any of these ethnic categories and were considered as "undetermined." In addition, a number of people who were not in the city when the censuses were conducted were probably not registered in them (see Andrews 1989, 81). Therefore, there might be an underestimation of the total size of the population and of the number of people of Amerindian and African origin.

In the census of 1887, the influence of the massive European immigration was evident: of 433,375 inhabitants, 425,370 were "white" and only 8,005 Afro-Argentinian. In that census there is no information regarding the Indian population.

Between 1880 and 1930, a significant number of immigrants coming from Europe settled in Buenos Aires, which produced a significant demographic growth of the city. According to the census of 1869, 14 percent of Argentina's population lived in Buenos Aires; by 1914, it had increased to 26 percent (Estrada and Salinas Meza 1987, 7), and 40–50 percent of its inhabitants were foreigners, mostly Italian and Spanish. Of a total of 3,500,000 immigrants of European origin, approximately 800,000 settled in the city Buenos Aires. Therefore as expected, in the middle of the twentieth century, the frequencies of ABO and Rh systems were similar to those of Spain and Italy (Avena 2003, 48). In 1936, even though the European immigration had begun to decrease, 36 percent of the BAMA population was immigrant (Dirección General de Estadística y Censos de la Ciudad de Buenos Aires 2009, 103). As from 1940, the industrial development of the city attracted immigrants of the provinces and also of the bordering countries, which have a high Hispano-Amerindian composition and a low African one (Avena 2003, 67).

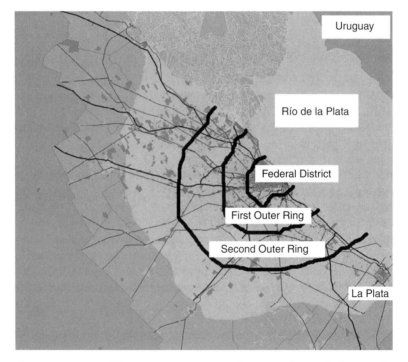

Figure 8.1   Map of the Buenos Aires Metropolitan Area (BAMA), showing the three areas (city of Buenos Aires, first outer ring, and second outer ring).

In the city of Buenos Aires in 1947, 94 percent and 6 percent of the foreigners came from European and South American countries, respectively (Dirección Nacional de Servicio Estadístico 1947, 59). On the other hand, in 2001, those values were 34 percent and 66 percent, respectively (Instituto Nacional de Estadística y Censos 2001). While the number of inhabitants of the city remained fairly constant (3,090,000 inhabitants, census 2001), the population of the outskirts grew fivefold between 1947 and 2001 from 1,700,000 to 8,000,000 inhabitants. The second outer-ring suburb was settled largely by migrants from the provinces and the bordering countries. Consequently, this movement of people modified the ethnic composition of the BAMA population (see the distribution of this area in figure 8.1).

Methods

The samples studied were obtained from the Hospital Italiano and Hospital de Clínicas of the city of Buenos Aires. The latter is a public hospital, and the former used to be a health centre that treated, during the first half of the twentieth century, members of the Italian community in Argentina, but it nowadays attends people of different ethnic origins who live in Buenos Aires and other parts of the country. Avena et al. (2006, 115) have demonstrated that there is no significant difference in both samples regarding the protein- and molecular-gene frequencies; so for our present study, we joined and treated them as if they were only one sample.

This study was performed on 211 unrelated individuals. The presence of the C-T transition in the DYS 199 locus was determined according to the protocol of Underhill et al. (1996, 115). In South Amerindians, this transition has a frequency of 85–90 percent and is absent in European and African people (Bianchi et al. 1997, 86).

The mtDNA was examined by the Restriction Fragments Length Polymorphisms (RFLPs) at three sites and in the region V 9-bp deletion (8272–8289), which allowed the identification of the four Amerindian haplogroups: haplogroup A is defined by the gain of a HaeIII restriction site at nt position 663; haplogroup B by the presence of the 9-bp deletion; haplogroup C by the absence of a HincII 13259 site, and haplogroup D by the absence of an AluI 5176 site. A positive HpaI site at nt 3592 was used to identify the mtDNA lineage of African origin L1 y L2.

For the identification of Amerindian and sub-Saharan mitochondrial haplogroups, five sequences for each individual were amplified by Polymerase Chain Reaction (PCR) in an independent way, following protocols already described by Stone and Stoneking (1993, 465).

European lineages were identified by at least one RFLP. In the case of W/I haplogroups, we only assay the RFLP that is shared by both, and they were not further analyzed. Sequencing of the hipervariable region I (HVRI) was done to confirm the assignment of X haplogroup, because it is present in both European and North American natives. Details for primers, restriction points, and region sequenced will be provided by the authors when requested. Protocol for PCR was the same as described before.

The frequency of the haplogroups was determined by direct counting and a survey was performed to obtain genealogic information of the donors.

The sample was divided into three regions according to the donors' residence: first, the city of Buenos Aires, and second, outer-ring suburbs. An X^2 test was performed to evaluate differences between them. The second outer-ring suburb was expected to present the highest frequencies of Amerindian haplogroups, as this is an area inhabited by the majority of migrants from provinces, bordering countries, and Peru, who have a high Hispano-Amerindian composition.

It is interesting to mention that the sample analyzed in this paper was also studied by the determination of ten erythrocyte genetic systems. To estimate if it had some consistency, we compared the allele frequencies of ABO and Rh systems to those calculated from 11.883 donors registered in Hemotherapy Service of Hospital of Buenos Aires in 2001. The observed differences were not significant (ABO: $X^2 = 1.07$, P> 0.01 and Rh: $X^2 = 0.21$, P > 0.01) (Avena et al. 2006, 114).

Results

Of the 211 individuals studied, 102 (48 percent) presented Amerindian lineages. At regional level, according to the maternal grandmother's place of birth, we can observe that 66/102 (65 percent) of the Amerindian haplogroups came from the provinces: 23/102 (22 percent), 16/102 (16 percent), 21/102 (21 percent), and 6/102 (6 percent) from northwest (NWA), northeast (NEA) of Argentina, Central, and Cuyo/Patagonia regions, respectively. In addition, 23/102 (22 percent) of the donors had grandparents whose places of birth were located in bordering countries and Peru. Therefore, a total of 89/102 (87 percent) Amerindian haplogroups came from ancestors who were not born in the BAMA. This region contributed with only 10/102 (10 percent). According to the inquiry performed, two donors (2 percent) who had European grandmothers presented Amerindian haplogroups as well. These results could be the consequence of illegitimacy or the donors' misinformation regarding their grandmother's place of birth. One donor of Asian ancestry (1 percent) presented Amerindian haplogroup, and only one African lineage from the Central region of Argentina was detected (table 8.2A).

The donors whose grandparents were born in the BAMA region presented high frequencies of European lineages and together with those of European origin added up to 83/108 (77 percent). Inversely, the NWA, NEA, Cuyo-Patagonia regions, and the bordering countries and Peru contributed with only 7/108 (6 percent), and the Central region with 18/108 (17 percent) of those lineages. The total European haplogroups were distributed in the following way: 60/108 H (55 percent), 13/108 U (12 percent), 9/108 T (8 percent), 8/108 K (7 percent), 5/108 V (5 percent), 7/108 I-W (6 percent), 4/108 J (4 percent), and 2/108 X (2 percent). As expected, the two X haplogroup did not exhibit the typical Amerindian mutation in HVRI 16213G—A, so they were assigned as Europeans.

Tables 8.2B and 8.2C show how the haplogroups are distributed in different places in the BAMA (see figure 8.1). The residents of the city of Buenos Aires and those of the first outer-ring suburb have a similar proportion of matrilineal lineages; for this reason, we decided to organize the data in only one table.

In table 8.2B, we show that Amerindian mitochondrial lineages add up to 59/148 (40 percent), but the majority of these haplogroups came from NWA (11/59: 19 percent), NEA (10/59: 17 percent), Central (11/59:19 percent), Cuyo/Patagonia regions (4/59: 7 percent), and bordering countries and Peru (15/59: 25 percent). Only 7/59 (12 percent) are residents whose grandparents were born in the BAMA. One haplogroup came from people who were born in Asia (2 percent). In addition we can observe that 88/148 (59 percent) presented European haplogroups, of which 15/48 (31 percent) are from the BAMA region, 9/48 (19 percent) came from the provinces, and 24/48 (50 percent) are the donors whose grandparents were born in Europe. Lastly, we found only one African haplogroup.

If we analyze the data from the second outer-ring suburb, we can see that the frequency of Amerindian lineages is the highest of the two areas compared. In fact, these lineages add up to 43/63 (68 percent). The difference between the areas is significant ($X^2= 12.3$, $p < 0.001$). On the other hand, European maternal lineages are lower 20/63 (32 percent) (table 2C). These results were to be expected given that, as we have mentioned before, this area was settled by the majority of the migrants that came from the provinces and the bordering countries and Peru, which all have a high Hispano-Amerindian composition.

On the basis of this analysis, it is evident that the majority of the Amerindian mitochondrial haplogroups came from women who were

Table 8.2 Mitochondrial DNA of the donors and place of birth of maternal grandmothers

| | Haplogroups | | | | | | | | | | | | | | | | | | |
| | Amerindian | | | | Total Amerindian | % | European | | | | | | | | Total European | % | Total African | | Total |
Region / country	A	B	C	D			H	J	K	T	U	V	I-W	X			L	%	
(A) Donors and place of birth of maternal grandmothers																			
BAMA	2	0	5	3	10	24	16	1	1	2	3	3	4	1	31	76	0	0	41
NWA	5	3	9	6	23	96	1	0	0	0	0	0	0	0	1	4	0	0	24
NEA	8	4	3	1	16	89	0	0	1	0	1	0	0	0	2	11	0	0	18
Central	2	5	7	7	21	53	9	1	2	4	2	0	0	0	18	45	1	3	40
Cuyo/Patagonia	2	1	2	1	6	75	2	0	0	0	0	0	0	0	2	25	0	0	8
Italy	1	0	0	0	1	4	14	2	2	2	2	1	2	1	26	96	0	0	27
Spain	0	0	0	1	1	8	8	0	1	0	1	1	0	0	11	92	0	0	12
Other European and near east	0	0	0	0	0	0	9	0	1	1	3	0	1	0	15	100	0	0	15
Bordering countries and Peru*	3	11	9	0	23	92	1	0	0	0	1	0	0	0	2	8	0	0	25
Asia (far east)	0	0	0	1	1	100	0	0	0	0	0	0	0	0	0	0	0	0	1
Total	**23**	**24**	**35**	**20**	**102**	**48**	**60**	**4**	**8**	**9**	**13**	**5**	**7**	**2**	**108**	**51**	**1**	**0.5**	**211**
(B) Residents of Buenos Aires and the first outer-ring suburb and place or birth of maternal grandmothers																			
BAMA	1	0	3	3	7	19	15	1	1	2	3	2	4	1	29	81	0	0	36
NWA	2	1	5	3	11	92	1	0	0	0	0	0	0	0	1	8	0	0	12
NEA	6	3	0	1	10	91	0	0	1	0	0	0	0	0	1	9	0	0	11
Central	1	4	2	4	11	50	6	1	1	1	1	0	0	0	10	45	1	5	22
Cuyo/Patagonia	2	1	0	1	4	67	2	0	0	0	0	0	0	0	2	33	0	0	6
Italy	0	0	0	0	0	0	12	2	2	2	2	0	2	1	23	100	0	0	23

	1	2	3	4	5	6	7	8	9	10	11	12	13	14	15	16	Total
Spain	0	0	0	0	0	0	6	1	0	1	1	0	0	9	100	0	9
Other European and near east	0	6	0	0	0	0	6	1	1	3	0	1	0	12	100	0	12
Bordering countries and Peru**	2	0	7	0	15	0	0	0	0	1	0	0	0	1	6	0	16
Asia (far east)	0	0	0	1	1	100	0	0	0	0	0	0	0	0	0	0	1
Total	**14**	**15**	**17**	**13**	**59**	**40**	**48**	**7**	**6**	**11**	**3**	**7**	**2**	**88**	**59**	**1**	**148**
(C) Residents of outskirts—second and third outer-ring suburb—and place of birth of maternal grandmother																	
BAMA	1	0	2	0	3	60	1	0	0	0	0	0		2	40	0	5
NWA	3	2	4	3	12	100	0	0	0	0	0	0		0	0	0	12
NEA	2	1	3	0	6	86	0	0	1	1	1	0		1	14	0	7
Central	1	1	5	3	10	56	3	1	3	1	0	0		8	44	0	18
Cuyo/Patagonia	0	0	2	0	2	100	0	0	0	0	0	0		0	0	0	2
Italy	1	0	0	0	1	25	2	0	0	1	0	1		3	75	0	4
Spain	0	0	0	1	1	33	2	0	0	0	0	0		2	67	0	3
Other European and near east	0	0	0	0	0	0	3	0	0	0	0	0		3	100	0	3
Bordering countries and Peru***	1	5	2	0	8	85	1	0	0	0	0	0		1	11	0	9
Total	**9**	**9**	**18**	**7**	**43**	**68**	**12**	**1**	**3**	**2**	**2**	**1**		**20**	**32**	**0**	**63**

Notes: BAMA = Buenos Aires Metropolitan Area; NWA = Northwest of Argentina; NEA = Northeast of Argentina; Central = Central Region of Argentina

* People come from Bolivia (3), Brasil (1), Chile (1), Paraguay (14), Peru (1), and Uruguay (5)

** People come from Bolivia (3), Brasil (1), Chile (1), Paraguay (6), Peru (1), and Uruguay (4)

*** People come from Uruguay (1) and Paraguay (8). Sub-Saharian mitochondrial haplogroups was not detected.

born outside the BAMA region. Moreover, as was expected in South American cosmopolitan populations, Amerindian paternal lineages (4 percent) were very low in relation to maternal ones.

When we analyze the types of marriage between grandparents of the donors, we observed a very strong endogamic trend (table 8.3). In the diagonal of this table that goes from top to bottom we can see this characteristic clearly. The percentage of endogamic marriages of the grandparents has a range of variation from 77 percent to 90 percent and from 75 percent to 100 percent for females and males, respectively. In this analysis we have not taken into account the Asian people because of their small size.

Besides, we have also analyzed all the grandparents' marriages including those whose descendants were not studied in the laboratory. The endogamic trend is very strong both in the female and male ancestors of the donors, with a range of variation from 74 percent to 100 percent and 69 percent to 100 percent, respectively. Likewise, it is significant to point out that the majority of European women were married to European males (90 percent). The same tendency can also be observed among the other geographic groups; for example, 100 percent of the females of the NEA were married to males of the same origin. This high endogamy, according to the geographic origin of the partners, suggests that these biological unions could be determined by economic, cultural, and social barriers.

Table 8.3   Types of marriage between the maternal grandparents of donors

Female \ Male	BAMA	NWA / NEA	Central	Cuyo / Patagonia	European countries	South American countries	Asia	Total	Female endogamic percentage
BAMA	31	–	1	–	7	1	–	40	77
NWA / NEA	–	29	1	–	2	–	–	32	90
Central	2	1	27	–	3	–	–	33	82
Cuyo / Patagonia	–	–	–	7	1	–	–	8	87
European countries	4	1	3	–	42	–	–	50	84
South American countries	1	1	1	–	1	15	–	19	79
Asia	1	–	–	–	–	–	4	5	80
Total	39	32	33	7	56	16	4	187	–
Male endogamic percentage	79	91	82	100	75	94	100	–	–

Discussion

In Argentina, two historic events had a strong influence in its genetic composition. The first one was the massive European immigration from 1880 to 1930, when almost 900,000 people inhabited the city, and its population presented similar frequencies of ABO and Rh systems to Italy and Spain. The second took place in the 1940s, when the industrial development attracted people from the provinces and the bordering countries, with a high Hispano-Amerindian composition. This last migration modified the gene pool of the BAMA population.

The results of our study, from the analysis of the uniparental genetic markers, seem to confirm this observation. Forty-eight percent, 51 percent, and 0.5 percent of Amerindians, European, and African maternal lineages were observed, respectively, but they presented a differential distribution according to the BAMA regions. The residents of the city of Buenos Aires and the first outer-ring suburb presented a similar average percentage (40 percent) of Amerindian maternal lineages, against 68 percent of the second outer-ring suburb. These data are in accordance with the fact that people who came from the provinces, the bordering countries, and Peru settled in that area. Conversely, the European maternal lineages presented lower percentages (32 percent) than those in the city of Buenos Aires and the first outer-ring suburb (59 percent). As we mention before, we studied the same sample through the determination of protein markers, and we detected minor values of Amerindian gene admixture (16 percent) and major values (3.3–4.0 percent) of African component (Avena et al. 2001, 86; Avena et al. 2006, 115). These differences between uniparental- and biparental-genetic markers were also observed in a sample of Bahía Blanca, province of Buenos Aires. In this study, the authors detected 47 percent of Amerindian maternal lineages, and this percentage fell to 36.8 percent using autosomal Alu polymorphic insertion (Avena et al. 2006, 66; Resano et al. 2007, 833). In turn, V. Martinez-Marignac et al. (1999, 294; 2004, 543) in the city of La Plata, province of Buenos Aires, detected 50 percent of Amerindian maternal lineages, and this figure decreased to 26 percent when they used autosomal molecular markers. The same trend was observed in Comodoro Rivadavia in Patagonia region, where 70 percent of indigenous maternal haplogroups was detected, whereas the gene admixture from biparental genetic markers was 37 percent (Avena et al 2009, 32). Finally, D. Corach et al. (2010) in three Argentinian regions also observed higher frequencies of Amerindian maternal lineages (53.7 percent) than those revealed by autosomal DNA (17.3 percent).

It is interesting to compare these data with those observed in Uruguay, where S. Pagano et al. (2005, 1240, modified in Sans 2009, 167) detected 31 percent of Amerindian maternal ancestry, and gene admixture estimated from nuclear markers was 10 percent (Hidalgo et al. 2005, 218). Moreover, in samples of Santiago de Chile, P. Rocco et al. (2002, 127) recorded 84 percent of Native American maternal contribution, though, when the gene admixture was estimated by using nuclear genetic markers the level of indigenous participation was 40 percent.

Therefore, these different values of genetic admixture may be the result of the different paths that the gene transmission followed, which in the case of biparental markers both parents contribute to the mixture process (see Martínez-Marignac et al. 2004, 546). With respect to the contribution of African maternal lineages, population samples from Argentina showed a variation range from 2 percent to 6 percent percent, while in Uruguay the range varied according to region, from 8 percent to 21 percent (Bonilla et al. 2004, 291, Pagano et al. 2005, 1240, modified in Sans 2009, 167 and Sans et al. 2006, 519).

On the other hand, we observed significant differences between the Amerindian maternal and paternal lineages, given that only 4 percent of the donors presented DYS199T allele. Recently, we analyzed 17 Y short tandem repeats (STRs) in 85 males of this sample. Ninety-three percent of the haplotypes could be assigned mainly to European paternal lineages; haplogroup R1b (39 percent) was the most frequent, followed by G (11 percent), I (10 percent), J (9 percent), and R1a (6 percent). Other haplotypes, such as E3b (15 percent), K2 (2 percent), and N (1 percent) present in people of European origin, were also detected in Eurasia, North Africa, and East Asia, respectively. The Q3 Amerindian paternal lineage presented low frequency (6 percent), and all presented the haplotype markers Q3 described in Amerindian populations (DYS19*13, DYS390*24, DYS391*10, DYS392*14, and DYS393*13). Only one haplogroup was of African origin, E3a (1 percent). In addition, the minimum haplotypes of nine loci STRs were analyzed (DYS19, DYS389I, DYS389II, DYS390, DYS391, DYS392, DYS393, and DYS385a/b) and 88 haplotypes were detected, 21 percent of them did not match those of the Y Chromosome Haplotype Reference Database (YHRD) (YHRD, $n = 72946$), while the rest (79 percent) were of European origin, mainly from Italy and Spain (data not published). These results are in accordance with the percentages obtained in the present study, through the use of only DYS199 locus, as regards the low presence of Amerindian paternal lineages (APL) in the BAMA sample.

The asymmetry by sexes was also observed in other populations of Argentina. J. E. Dipierri et al. (1998, 1007) in a sample of San Salvador de Jujuy, northwestern Argentina, detected 100 percent of Amerindian maternal lineages (AML) compared with 36 percent of APL. This feature is also presented in the northern province of Cordoba (AML = 79 percent and APL = 8 percent, García and Demarchi et al. 2006, 65), in Bahia Blanca in southern Buenos Aires Province (AML = 47 percent and APL = 4 percent, Avena et al. 2007, 66–68), and Comodoro Rivadavia in Patagonia region (AML = 70 percent and APL = 6 percent, Avena et al. 2009, 32–34). The same trend was observed in other South American countries. In the city of Belem, located in the northeastern part of the Brazilian Amazon region, Batista dos Santos et al. (1999, 177) observed, also, different contributions of maternal and paternal haplogroups (AML = 50 percent, and APL = 4 percent, and 3 percent of AML). In turn, based on analysis of 52 individuals from the cities of Pasco and Lima, Peru, Rodriguez-Delfin et al. (2001, 97) commented that "there is a clear directionality of marriage, with an estimated genetic admixture with non-Amerindians that is 9 times lower for mtDNA than for Y chromosome DNA."

In Uruguay, M. Sans et al. (2002, 36), in a sample of African descendants in the town of Melo, could prove a differential indigenous maternal and paternal contribution to the gene pool of that population (AML = 29 percent and APL = 6 percent). Also, in Antoquian in the northwest of Colombia was a clear sex differential contribution (AML = 90 percent and APL = 1 percent), while the European and APL represented 94 percent and 5 percent, respectively (Bedoya et al. 2006).

On the other hand, in spite of the significant presence of African people in the mid-eighteenth century we detected only one African maternal lineage, which represented less than 1 percent of the sample. However, as mentioned in the introduction of this chapter, we studied the same sample with biparental genetic markers and observed 2.2 percent and 4.3 percent of African contribution (Fejerman et al. 2005, 168). In addition, Fejerman et al. 2005 (168–169), observed that the individual admixture showed that those alleles that have a high frequency in populations of African origin tend to concentrate among eight individuals. Therefore, although the admixture estimate is relatively low, the actual proportion of individuals with at least some African influence is approximately 10 percent. Anyway, the causes of phenotype "invisibility" of the African components in the BAMA are not easy to explain not only because of their significant presence in the eighteenth century in the city of Buenos Aires, but also for their high presence in northern Argentina (see Lorandi 1992, 157–158), which was the region that provided migrants to the BAMA. Some authors believe that the "disappearance" of the African people could have several causes such as: (1) the abolition of the slave trade; (2) the high male mortality produced during several wars in the eighteenth century, especially the war against Paraguay (1864–1870); (3) high mortality and low fecundity among them; (4) the yellow fever epidemic in Buenos Aires of 1871; and (5) the "dilution" as a consequence of the admixture with people of different origins (see Andrews 1989, 10–12). We believe that this last hypothesis could be the most important cause of the phenotype "disappearance" of African people, given that in the nineteenth century they made up a quarter of the 60,000 inhabitants of the city, which then received, approximately, 800,000 European immigrants at the end of that century; in addition, the low proportion of African males allowed the biological unions between African females with males that came from Europe (Andrews 1989, 10). In these cases, we expected to find a higher frequency of African maternal haplogroups; however, the low frequency detected may be explained by the loss of maternal lineages in some of the generations of the donors' ancestors, which would also explain the different percentage of mixture obtained regarding biparental markers.

So far, we have analyzed how the European, Amerindian, and African lineages are distributed in the BAMA. Regarding the types of marriages, the maternal grandparents' marriages of the donors showed a strong endogamic trend. If we divide the marriages by the donors' origin, we see that the Europeans and their descendants show a high endogamic behavior. The same tendency was also observed in all the other geographic groups (table 8.3). When we analyzed all the grandparents' marriages, 100 percent, 90 percent, and 92 percent of the females from NEA, Europe, and the bordering countries, respectively, were married to partners of the same regions.

On the other hand, we estimate in 35 the average age of the donors and in 25 each generation, therefore the donors grandparents were born approximately in the 1920s. Until that time, as we can see above, the European marriages were mainly endogamic. As to this issue, it is interesting to ask: What happened to the European marriage behavior before the time analyzed in this study? In a previous study, based on data from the Registers of Births, Marriages, and Deaths of the city of Buenos Aires, we analyzed the types of marriages of Spanish immigrants and their descendants from 1890 to 1900, and we observed the same endogamic trend (Caratini et al. 1996, 66–6 9). In this article, from a total of 1,100 marriages analyzed, they noted 56 percent and 44 percent of endogamic and exogamic marriages, respectively; however, of the 44 percent of the exogamic marriages (Argentinian males married to Spanish females), 32 percent of the males were of Spanish origin, 17 percent of Italian origin, and 10 percent of French origin. The majority of the other marriages (41 percent), whose members were born in Argentina, had surnames of Italian and Spanish origins. In addition, it is interesting to mention that the most important migratory current in Argentina was of Italian origin

(Devoto 2003, 247). Some authors have studied the pattern of marriages of these immigrants from the eighteenth to the nineteenth century, and they have also observed a high endogamic tendency among them (Seefeld 1986, 214–223).

The analysis of the types of marriages has demonstrated that the Europeans had a low admixture with people who were born in different regions of Argentina, and also with other people of South American countries. This high endogamic behavior was observed also in all the groups analyzed.

Based on the data analyzed so far, we can conclude that: (1) the majority of donors whose grandparents were born in provinces and the bordering countries have a high Amerindian composition, and they were the ones who carried the maternal haplogroups inherited from their ancestors to the BAMA. This conclusion is in accordance with the fact that 83 percent of Amerindian maternal lineages came from those regions. In turn, as we can see in table 8.3, the grandparents of those regions in general did not mix with partners from the BAMA or from European origin; (2) the low proportion of indigenous populations in the city of Buenos Aires, according to the censuses from 1806 to 1827 (table 8.1), made us think that the probability of marriage with European partners was null or minimum; and (3) the high endogamy observed in the preceding generations of the donor's, added to the data mentioned in (1) and (2) suggests that the gene admixture between Amerindians and non-Amerindians could have taken place before that time, probably during the colonial period and mainly in the NWA-NEA and Patagonian regions. Therefore, the focus given to this study allowed us to observe that the current genetic composition, according to the sample analyzed, of the BAMA was not a consequence of mixed marriages in this area, but a consequence of people already mixed in their places of origin. Consequently, based on the genetic and genealogical data obtained, we can postulate that the "melting pot" has not occurred to a great extent in the BAMA. However, we are aware that the sample studied is small, and therefore more studies will be necessary to reach conclusive results.

Acknowledgments

We would like to thank the authorities of the Hospital Italiano and the Hospital de Clínicas of the city of Buenos Aires for backing this study. We would also like to thank the blood donors, who were adequately informed about the aims of the study and who gave their approval to carry out this research. Financial support was provided by the Consejo Nacional de

Investigaciones Científicas y Técnicas (CONICET) and the Secretaría de Ciencia y Técnica de la Universidad de Buenos Aires.

References

Andrews, G. R. (1989) *Los Afroargentinos de Buenos Aires*. Buenos Aires: Ediciones de la Flor.

Avena, S. A. (2003) Análisis antropogenético de los aportes Indígena y Africano en muestras hospitalarias de la Ciudad de Buenos Aires. Unpublished thesis (PhD), Universidad de Buenos Aires.

Avena, S. A. et al. (1999) Análisis de la participación del componente indígena en una muestra hospitalaria de la ciudad de Buenos Aires. *Revista Argentina de Antropología Biológica* 2(1), pp. 211–226.

Avena, S. A. et al. (2001) Análisis antropogenético de los aportes indígena y africano en muestras hospitalarias de la ciudad de Buenos Aires. *Revista Argentina de Antropología Biológica* 3(1), pp. 79–99.

Avena, S. A. et al. (2006) Mezcla génica en la Región Metropolitana de Buenos Aires. *Medicina* 66, pp. 113–118.

Avena, S. A. et al. (2007) Mestizaje en el sur de la región pampeana (Argentina). Su estimación mediante el análisis de marcadores proteicos y moleculares uniparentales. *Revista Argentina de Antropología Biológica,* 9(2), pp. 59–76.

Avena, S. A. et al. (2009) Mezcla génica y linajes uniparentales en Comodoro Rivadavia (Prov. de Chubut, Argentina*). Revista Argentina de Antropología Biológica,* 11(1), pp. 25–41.

Batista dos Santos, S. E. et al. (1999) Differential contribution of indigenous men and women to the formation of an urban population in the Amazon region as revealed by mtDNA and Y-DNA. *American Journal of Physical Anthropology,* 109(2), pp. 175–80.

Bedoya, G. et al. (2006) Admixture dynamics in Hispanics: A shift in the nuclear genetic ancestry of a South American population isolate. *Proceedings of the National Academy of Science (U. S. A.),* 103(19), pp. 7234–7239.

Bianchi, N. O. et al. (1997) Origin of Amerindian Y-chromosomes as inferred by the analysis of six polimorphic markers. *American Journal of Physical Anthropology,* 102, pp. 79–89.

Bonilla, C. et al. (2004) Substantial Native American ancestry in the population of Tacuarembó, Uruguay, detected using mitochondrial DNA polymorphisms. *American Journal of Human Biology,* 16, pp. 289–297.

Caratini, A. L. et al. (1996) Endogamia-exogamia grupal de los inmigrantes españoles en la ciudad de Buenos Aires: Su variación en el espacio y en el tiempo. *Revista Española de Antropología Biológica,* 17, 63–75.

Corach, D. et al. (2010) Infering continental ancestry of Argentineans from autosomal, Y-chromosomal and mitochondrial DNA. *Annals of Human Genetics,* 74, 65–76.

Devoto, F. (2003) *Historia de la inmigración en la Argentina*. Buenos Aires: Sudamericana.

Dipierri, J. E. et al. (1998) Paternal directional mating in two Amerindian sub-populations located at different altitudes in northwestern Argentina. *Human Biology*, 70(6), pp. 1001–1010.

Dirección General de Estadística y Censos de la Ciudad de Buenos Aires (2009) *El Censo de 1936. Cuarto Censo General de la Ciudad de Buenos Aires. Población de Buenos Aires.* Buenos Aires: Municipalidad de la Ciudad de Buenos Aires, 6, pp. 103–121.

Dirección Nacional del Servicio Estadístico (1947) *IV Censo Nacional de Población y Vivienda.* Buenos Aires: Presidencia de la Nación, 1, p. 59.

Estrada, B. and Salinas Meza, R. (1987) Inmigración europea y movilidad social en los centros urbanos de América Latina (1880–1920). *Estudios Migratorios Latinoamericanos*, 5, pp. 3–27.

Etcheverry, M. A. (1949) Frecuencia de los tipos sanguíneos Rh en la población de Buenos Aires. *Revista de la Sociedad Argentina de Hematología y Hemoterapia*, 1, pp. 166–168.

Fejerman, L. et al. (2005) The African ancestry of the population of Buenos Aires. *American Journal of Physical Anthropology*, 128(1), pp. 164–170.

García, A. and D. Demarchi. (2006) Linajes parentales amerindios en poblaciones del norte de Córdoba. *Revista Argentina de Antropología Biológica*, 8(1), pp. 57–71.

Hidalgo, P. C. et al. (2005) Genetic admixture estimate in the Uruguayan population based on the loci LDLR, GYPA, HBGG, GC and D7S8. *International Journal of Human Genetics*, 5, pp. 217–222.

Instituto Nacional de Estadística y Censos (2001) *Censo Nacional de Población y Vivienda del año 2001.* Ministerio de Economía, Presidencia de la Nación. Available from http://www.indec.gov.ar/censo2001s2_2/ampliada_index.asp?mode=02 [Accessed July 06, 2010].

Lorandi, A. M. (1992) El mestizaje interétnico en el noroeste argentino. *Senri Ethnological Studies*, 33, pp. 133–166.

Martínez-Marignac, V. et al. (1999) Estudio del ADN mitocondrial de una muestra de la ciudad de La Plata. *Revista Argentina de Antropología Biológica*, 2(1), pp. 281–300.

Martínez-Marignac, V. et al. (2004) Characterization of admixture in an urban sample from Buenos Aires, Argentina, using uniparentally and biparentally inherited genetic markers. *Human Biology*, 76(4), pp. 543–557.

Pagano, S. et al. (2005) A assessment of HV1 and HV2 mtDNA variation for forensic purposes in an Uruguayan population sample. *Journal of Forensic Sciences*, 50, pp. 1239–1244.

Resano, M. et al. (2007) How many populations set foot through the Patagonian door? Genetic composition of the current population of Bahía Blanca (Argentina) based on data from 19 Alu polymorphisms. *American Journal of Human Biology*, 19(6), pp. 827–835.

Rocco, P. et al. (2002) Composición genética de la población chilena: Distribución de polimorfismos de DNA mitocondrial en grupos originarios y en la población mixta de Santiago. *Revista Médica de Chile*, 130 (2), pp. 125–131.

Rodriguez-Delfín, L. A. et al. (2001) Genetic diversity in an Andean population from Peru and regional migration patterns of Amerindians in South America: Data from Y chromosome and mitochondrial DNA. *Human Heredity,* 51(1–2), pp. 97–106.

Sans, M. (2009) "Raza," adscripción étnica y genética en Uruguay. *RUNA,* 30(2), pp. 163–174.

Sans, M. et al. (2002) Unequal contributions of male and female gene pools from parental populations in the African descendants of the city of Melo, Uruguay. *American Journal of Physical Anthropology,* 118(1), pp. 33–44.

Sans, M. et al. (2006) Population structure and admixture in Cerro Largo, Uruguay, based on blood markers and mitochondrial DNA polymorphisms. *American Journal Human Biology,* 18, pp. 513–524.

Seefeld, R. F. (1986) La integración social de los extranjeros en Buenos Aires según sus pautas matrimoniales: ¿Pluralismo cultural o crisol de razas? (1860–1923). *Estudios Migratorios Latinoamericanos,* 1(2), pp. 203–231.

Stone, A. C. and Stoneking. M. (1993) Ancient DNA from a pre-Columbian Amerind population. *American Journal of Physical Anthropology,* 92, pp. 463–471.

Underhill, P. A. et al. (1996). A pre-Columbian Y chromosome-specific transition and its implications for human evolutionary history. *Proceedings of the National Academy of Science (U. S. A.),* 93, pp. 196–200.

9

National Identity, Census Data, and Genetics in Uruguay

Mónica Sans

As is found in all American countries, Uruguay's population is based on unequal proportions of Native Indians, Europeans, and Africans. The manner in which Uruguayans perceive these contributions has varied during different time periods and by different segments of the society, and is related in a broad sense to the concept of national identity. National identity can be defined by answering the questions: Who are we? What do we want? (Rodríguez Kauth, 2009). From this perspective, until at least two decades ago, two concepts were emphasized in Uruguayan national identity: the extermination of the Native population and the minimal contribution of Africans to national identity. The same concepts appear reflected in Brazilian anthropologist D. Ribeiro's (1969) work. In his historical and cultural configurations, he classified Uruguay as part of the "transplanted historico-cultural configuration," based on the fact that its population came almost exclusively from Europe and the presence of a European-based culture. Despite this, some African-related traditions are still very visible in Uruguay, especially during carnival, while Native American ones are mainly related to common use of Guaraní toponyms and the widespread use of *mate* (infusion of *Ilex paraguariensis*). More recently, some possible Charrúa and/or Native traditions have appeared, but it is not clear if they are traditions or recent inventions. For instance, some Uruguayan musicians have declared that they are trying to recreate *Charrúa* music, taking elements from nature, as the music might have been originally. In this paper, the historic process of peopling from Europe, Africa, and other regions, as well as Uruguay's relation to the Native inhabitants will be examined and analyzed in relation to national identity, recent census data involving "race" self-ascription, and genetic data.

Some Historical Aspects

At the time of the conquest, A. Rosenblat (1954) estimated that the present Uruguayan territory was populated by between 5,000 and 10,000 Native Indians, while J. H. Steward (1946) estimated the number to be closer to 20,000 including a broader region. M. Consens (2003), while identifying the population of Native Indians to be 5,000 individuals following events associated with conquest and colonization (disease, displacements), calculates ten-times more individuals living in some moments during the prehistoric times.

The territory was inhabited by three to four ethnic groups (Cabrera, 1992). The midland was inhabited by the macroethnic group Charrúa that included Guenoas, Bohanes, Yaros, and the largest, the Charrúa themselves. Yaros and Bohanes seem to have disappeared or to have integrated early, and soon after, the Guenoas (so called Minuanes) as well. Several documents mention hundreds or more Charrúas at different times: In 1574, Garay fought against more than 1,000 Charrúa Indians near San Salvador. In 1829, 200 warriors participated in the "Ejército del Norte," and in 1831, 1,000 were present in the massacre at Salsipuedes (Acosta y Lara, 1985; Sans and Pollero, 1991). The founding of San Francisco de Olivares de los Charrúas in 1624 was an attempt to put Charrúa Indians in *reducciones* (settlements founded by the colonizers to assimilate Native populations into the Spanish culture), but it lasted only for a short time (Rela, 2000). The Charrúa had a notorious presence during the first part of the nineteenth century, participating in the wars of independence from Spain. However, soon after, they were pursued and massacred during the first years after the founding of the republic. In 1831, near Salsipuedes in the northeast of the territory, a major offensive was carried out by the Uruguayan government with the aim of exterminating the Charrúa. President Rivera asked the Indians to meet him there to decide different aspects of their future, but instead Uruguayan troops killed most of the Charrúa men, while the women and children were captured (Acosta y Lara, 1969). Some references mention the *repartos* (distributions) of Charrúa women and children among Spanish or Criollo (Creole, Spanish, or Portuguese descendants born in Uruguayan territory) families in different regions. This was carried out after conflicts (wars or killings), during a relatively long period of time, but especially in the 1830s (among others, Acosta y Lara, 1969, 1985; Cabrera, 1983; Padrón Favre, 1986).

The second Native group, the Chaná, were located in the Argentinean border, near the Uruguay river. The last mention of them is probably related to the founding of the *reducción* of Santo Domingo de Soriano. This was initially in Argentinean territory but later moved to Uruguay between 1662 and 1664. This *reducción* also took in other Indians, and later, Spanish and Creoles. Again, some hundreds of Indians, in this case Chaná, are mentioned, but not until the seventeenth century (Acosta y Lara, 1989; Barreto, 2007). The third group was the Arachán; whose presence in the eastern region is not well documented, and they could have been restricted to Brazilian territory. Finally, the last ethnic group was the Guaraní, which arrived from the north around the fourteenth century using the rivers for traveling into the region.

The Guaraní had a unique role because their population growth occurred in historic times. The Catholic church records mentioned 1,048 Indians and Paraguayans (assuming most of them were Indians from the Jesuit Missions) in a total of 4,735 (23 percent) baptisms registered in parishes from the rural areas between 1771 and 1811 (González Rissotto and Rodríguez Varese, 1994). During the destruction at that time of the Portuguese Colonia del Sacramento (1680), 3,000 Guaraní Indians fought together with the Spaniards, and in the "Batalla del Yi" (1702), 2,000 supported the Spaniards against the Charrúa and other Indians (Sans and Pollero, 1991). In 1829 and 1832, two towns were founded in the northern part of the territory with around 8,000 Indians from the Jesuit Missions, most of them, probably Guaraní. These were not successful, in spite of the fact that one of them grew to eventually become the present town of Bella Unión. Little is known about what happened to the Guaraní Indians after these events, but most likely admixture with the overall population is the best explanation for their seeming disappearance. González Rissotto and Rodríguez Varese (1989) point out that their contribution has been essential to the formation of the Uruguayan society, because Guaraní or "Misioneros" (Indians from the Jesuit Missions) primarily living in the countryside started working in agriculture and collaborated in the foundation of towns. According to the authors, the admixture process between Iberians or Creoles and Indians, mainly from Guaraní origin, began very early, during the sixteenth century, and took place mainly in the rural areas in what is now Paraguay, Brazil, Argentina, and Uruguay. J. E. Pivel Devoto (1959) also stresses the importance of the Guaraní contribution, but he also mentions the incorporation of other Indians, as well as *gauchos* (usually, mixed Indian and Creole people living in the countryside), and African or African descendants.

Africans also made an important contribution to the population during historic times. There are references as early as the beginning of the Iberian colonization. Slaves are mentioned as taking part in the Portuguese founding of Colonia del Sacramento in 1680, although officially, slavery started after 1743 (Martínez Moreno, 1941). Africans came mainly from the Congo and Angola, as well as secondly from Ghana and surroundings. However, as almost half of these slaves came indirectly, having disembarked in Brazilian ports, the origin of some of them is difficult to establish (Isola, 1975). In 1830, the slave trade officially came to the end, but the illegal introduction of slaves continued for some years more in a number of complex ways, such as considering Africans, as well as fugitive slaves, as *colons* and therefore subjected to work under *tutela* (custody) until reaching the age of 25 years or during three years (Frega et al., 2004). Africans and their descendants continued entering Uruguay after the abolition of slavery (laws promulgated in 1841 and 1846, during the Guerra Grande War), mainly escaping from the system of slavery in Brazil, which continued until 1888. C. Rama (1967) mentions that the contribution of Brazilian blacks was very important throughout that period and also after, as they participated in the rice and sugarcane harvest as well as cattle raising, particularly in the northern region of the country, near the Brazilian border.

It is interesting to note that, at least during the second half of the nineteenth century, Negros (African or African descendants) were more visible in Uruguay than Indians or their descendants, and they are mentioned in religious festivities, guilds, schools, and different kinds of work. For instance, in 1853, information regarding domestic work in Montevideo, Uruguay's capital city, mentions that of 2,031 domestic workers, 76 percent were *de color* (colored) (Stalla, 2007; Frega et al., 2004).

The European populations that migrated to the territory are however better known and remembered by Uruguayan society than the Native and African populations. After the initial period of colonization by the Iberians, there were additional specific waves of migrants after independence (1825–1830); this includes French (mainly Basques), Brazilian, Spanish, Italian, and cosmopolitan (from different European and Mediterranean origins) (Pi Hugarte and Vidart, 1969).

Census Data

Before analyzing the census data in Uruguay, some further clarification is needed. Some census data refer to origin, usually the main three groups mentioned before (Indians or Natives, mostly referred as *Misioneros*, Europeans, and Africans). The emphasis has usually been on recording "race," usually understood as color (*blancos, negros, indios*). In this sense, during the eighteenth and nineteenth centuries, race can be seen as a mixed concept that usually combines color with social position, for example, a black slave, a free black, or a free or slave mulatto, or alternately only denotes color. In the twentieth and twenty-first centuries, in the census classification race is mixed with ethnic group (Instituto Nacional de Estadística, 2006), and races are joined for analysis. For instances in this case "black race" includes black and any mix with black, "Indian" includes Indian and Indian/white, with "white" being the only nonmixed group (Instituto Nacional de Estadística, 1997, 2006).

It is interesting to analyze C. Wagley's (1971) definition of social race, which is probably reflected in Uruguayan census material regarding race. The author stresses the heterogeneities found in the Americas, as a consequence of the characteristics of geography, of the Native populations, the conquerors, and the colonizers. He classifies the American countries in three different groups relating to the predominant origin of their inhabitants. In this analysis, Uruguay, together with Argentina, the United States, and Canada, belongs to the Euro-American group, mainly inhabited by European descendants. Wagley mentions that, in those countries, the criteria to classify social race is basically genealogical and related to the existence of nonwhite ancestors. However, in Uruguay, different than the United States, there are some intermediate categories, such as mulatto or *Pardo* (usually, mixed black and white), as well as mestizo (Indian and white). Consequently, in Uruguay as in other places in Latin America, the concept of race seems to be more related to color than to genealogy. This is a situation that stands opposed to the hypodescent concept that operates in the United States in which children of mixed unions between members of different ethnic groups are automatically assigned to the group that is considered subordinate or "inferior" and not to an intermediate or different group.

Only few Uruguayan census data sets have included race or population origin. In 1793, the country was populated by 30,885 inhabitants, including 7,000 black or mulatto, 700 Indians, and the rest, whites. Probably, this data only includes Indians living in towns, which can be deduced by taking into account the number of Indians participating in battles or killed, as mentioned before. For the Banda Oriental, the territory that included not only today's Uruguay, but also a large portion of Rio Grande do Sul, now Brazil, F. Azara (1809) mentioned 45,000 people at the end of the eighteenth century, 15,245 of them in Montevideo and more than 12,000 in the Jesuit Missions, most Indians. In 1842, the last census in the nineteenth century that includes data about race, the total population in the present territory of Uruguay indicates 200,000 inhabitants, but only

9,000 were considered to be black, with no mention of Indians or any mixed populations (Carvalho Neto, 1965; Martínez Moreno, 1941).

Most of the census data only includes the capital city, Montevideo. In 1751, with 939 inhabitants, 141 (15 percent) were black slaves. By 1778, the population had grown to 9,298 inhabitants: 6,695 Spaniards, 1,386 black slaves, 562 free blacks, 538 free *Pardos*, and 117 Indians. That is, 27 percent of the inhabitants were black or mulatto, and only 2 percent, Indians. Two years after that, the quantity of blacks was duplicated, while the quantity of mulattoes and whites tripled: in a total of 10,153 inhabitants, 228 were Indians, 2,050 blacks, 603 mulattoes, and the rest, whites. In 1803, a new census showed 899 (19 percent) black slaves, 141 (3 percent) free blacks and mulattoes, 3,033 (65 percent) whites. There is no data about the rest of the population (13 percent). In 1810, 36 percent were considered black (Martínez Moreno, 1941; Rosenblat, 1954; Carvalho Neto, 1965; Campagna, 1990). In 1829, the percentage of blacks went down to 15 percent while in 1843 it was 19 percent (Campagna, 1990). In these last cases, the inclusion of mulattoes in the black category is not clear. In spite of these figures, it is clear that European immigration increased rapidly in the second half of the nineteenth century, primarily in the capital city, but also spreading throughout the entire country.

No census data included race during the twentieth century, with the exception of a Household Survey that was conducted at the end of the century, in 1996–1997. At the mid-century, A. Rosenblat (1954) estimates 90 percent white, 2 percent black or mulattoes, and 8 percent mestizos, contrary to the national identity that gives priority to African descendants rather than mestizos. Similar data is shown in different editions of *The World Almanac*; for example, in the year 2000 (Famighetti, 2000), it mentions 89 percent whites, 10 percent mestizos, and 1 percent mulattoes and blacks. The origin of these data is uncertain, but it is probably based on data from unpublished estimations of the Uruguayan Ildefonso Pereda Valdés.

The 1996–1997 Continuous Household Survey asked for the self-declaration of race, and in 2006, the survey asked persons if they had ancestors of determined "races or ethnic groups." The first survey, carried out in 40,000 households (around 130,000 persons) located in towns with more than 5,000 inhabitants (which represent 86 percent of the Uruguayan population), asked for the explicit self-declaration of the surveyed population regarding the race to which they considered themselves to belong. The results highlighted that the declaration was a reflection of the self-perception of a relationship to a racial group. The options were white, black, Indian, yellow, and any mix grouped with the main races (Instituto Nacional de Estadística, 1997). The results for the 1996–1997 survey indicated that 93.2 percent of the people self-defined as white, 5.9 percent as black (category included black and any mix with black), 0.4 percent Indian (Indian and Indian white), and 0.4 percent yellow (probably, from Asia) (Instituto Nacional de Estadística, 1997).

Ten years after this survey, another census was performed, including race, but this time asking about ancestry. This new survey was applied to a sample of 18,506 houses in urban, suburban, and also rural areas. In this case, instead of the indication of the race to which an individual belonged, the question was: "Do you know if you have ancestors from…?" using the same categories as in the other survey, but this time termed as "race or ethnic group" (Instituto Nacional de Estadística, 2006). In this survey, 9.1 percent of the people indicated as having ancestors who were "Afro" (African origin) or black, while the percentage with Indian ancestry reached 4.5 percent; 94.5 percent indicated white ancestry, and only 0.5 percent, yellow. These percentages add up to more than 100 percent due to the fact that some people indicated more than one ancestry. The results also showed no differences between Montevideo and the other Departments throughout Uruguay, or between urban areas and small towns, or urban and rural areas. An unexpected result was that self-declared nonwhite ancestry was a little higher in Montevideo than in the countryside (Instituto Nacional de Estadística, 2006).

The inclusion of race and/or ethnic group in censuses or surveys is not new in the Americas, although it was not common in Uruguay where no data regarding this aspect was registered between 1842 and 1996. The United Nations (2007) stresses that some countries have the need to identify ethnic or national groups guided by different criteria regarding race, color, language, or religion. The document advises that these categories should be determined by the groups to identify themselves. The inclusion of a race module in the 1996–1997 survey in Uruguay seems to follow this rationale. It was proposed by African descendants who were organized in the Mundo Afro association and was included in the Continuous Household Survey by the state's Instituto Nacional de Estadísticas (Saura, 2008).

It is necessary to stress that, in both surveys, as well as in previous censuses, the meaning of "race" is not related to biological races that refer, according to T. Dobzhansky's definition (1955), to populations that differ from one another in the frequency of certain genes and that have the possibility of exchanging genes across geographic barriers that usually separate them. The existence of human races has been debated from different perspectives (see, for example, Armelagos, 1995; Salzano, 1997; Long and Kittles, 2003). In Uruguay, censuses including color (or eventually, race) carried out during the eighteenth and nineteenth centuries seem to be more concerned about social and economic status (free or slave) than about the color/race itself. This can be supported by the fact that, after the abolition of slavery, no census included race until the survey done at the end of the twentieth century. Otherwise, the criteria used in Uruguay in the last surveys is related to population origins (African, Native, Asian, and European), and more closely related to Wagley's (1971) social race, as defined previously. In these cases, "race or ethnic group," are used synonymously. The equivalence between race and ethnic group (this last, used specifically in the 2006 survey) has been discussed previously by J. Huxley and A. C. Haddon (1936) and A. Montagu (1942), vindicating the use of the first related to biological aspects, and the second, to cultural ones. It was Montagu who proposed ethnicity to describe human subdivisions, instead of using race.

National Identity

C. Zubillaga (1992) defines national identity as based on different aspects: a sense of belonging, historical consciousness, cultural differentiation, political autonomy, traditions, economic possibilities, multiethnic coexistence, social projects, and the overcoming forms of injustice.

Up until the 1970s, there were few doubts about what constituted "Uruguayan identity," and the country had what might be described as an "arrogant" vision in comparison with the other Latin American countries. There was a widespread perception that Uruguay was a country whose population and culture were transplanted from Europe directly; it was characterized by its democratic values, together with an almost totally white population (Demasi, 1995). This vision can be clearly seen in an official book published to commemorate the centenary of Uruguayan independence, *El Libro del Centenario de 1825* (Ministerio de Instrucción Pública, 1925). The book stresses that Uruguay was the only American nation that can categorically assert that within its boundaries there was no population that remembered its aboriginal history, as well as an absence of the atavistic problems that divisions of race and religion have provoked in other countries. At that time (1925) national identity was based on the historic recounting of the massacre of Charrúa Indians in 1831, together with their disappearance as an ethnic group, the intentional ignoring of other Indian Natives in Uruguay, and the waves of European immigrants that occurred mainly during the second half of the nineteenth century. This kind of reading reflected the vision and perspective from the capital city, the recent immigration, the *Colorado* political party, and the opposition to a nationalist and rural view. There were other perspectives that paralleled this model regarding populations in other parts of the Americas, such as those previously mentioned by Ribeiro (1969) and Wagley (1971), or earlier by R. Bilden (1931).

In the decade of the 1980s and especially after the end of the dictatorship period (1973–1985), different analyses generated alternative perspectives on the "racial" composition of the Uruguayan population views. C. Demasi (1995) alludes to the recreation of the foundational retelling of the genocides, mainly in relation to the event at Salsipuedes. On the other hand, G. Verdesio (1992) emphasizes the deconstruction of the official history through the reanalysis of the colonial period and the reemergence of diversity and the voice of the subjugated (minorities). M. Viñar (1992) criticizes the image of Indians as domesticated, as well as their assumed quiet and silent disappearance at the beginning of the country's independence. Moreover, C. Aguiar (1992) stresses the fallacy of a vision of an empty country before European colonization, together with the illusion of an hyperintegrated society and the ethnic absorption of Indians.

Currently, Uruguay seems to have consolidated a somewhat new identity. During the FIFA South Africa World Cup 2010, different sources mentioned the *garra Charrúa*, defined as "the term Uruguayans use for that quality of gritty determination that any self-respecting Uruguay team, especially one preparing for the FIFA World Cup finals, must have in their DNA" (World Cup News, 2010). Curiously, the definition includes DNA, as a new interpretation of a cultural value. After multiple celebrations as a consequence of the Uruguayan fourth position in the World Cup soccer competition, Uruguayan president José Mujica declared, when greeting the returning soccer players at the Congress: *"Nunca hemos estado tan unidos por encima de las clases sociales y los colores políticos"* (we have never before been so united regarding social classes and political parties), posted in the newsletter (El País Digital, 2010).

Genetic Data

In parallel with the perspective regarding the Uruguayan national identity, most of the genetic studies until the decade of the 1980s were focused on samples of the "white" population and made comparisons with European countries, mainly focusing on Spain. In 1986, the study of a predominantly non-European trait, the Mongolian spot, showed values that clearly disagreed what might be expected from a white population (42 percent instead of the expected less than 10 percent)[1] (Sans et al. 1986). That study, together with other genetic data, began to challenge the idea of a "European transplanted" identity. It also opened a new era in populational genetics. This was initially based on autosomal-inherited genetic polymorphisms whose frequency differs among populations along with the application of methods to estimate population contributions. This allowed estimations to be made of the three main contributions to the Uruguayan population. The results showed not only higher African and, especially, higher Indian contributions than estimated based on assumed national identity at that time, but also, regional differences. For example, the Indian contribution was estimated as 20 percent in Tacuarembó, in the northeast of the country, but as only 1 percent in Montevideo, with regions such as Cerro Largo (8 percent) showing intermediate values (Sans et al., 1997, 2006). The only study published at present that includes a sample that represents the entire country, based on several nuclear DNA loci, reveals an Indian contribution of 10 percent (Hidalgo et al., 2005). By contrast, the African contribution seems to be more homogeneous in different regions of the country. The mentioned studies outlined above estimated this contribution to be 15 percent in Tacuarembó, 10 percent in Cerro Largo, and 7 percent in Montevideo, with 6 percent for the entire country.

However, the uniparental-inherited markers (mtDNA and Y chromosome) give a different perspective. Mitochondrial DNA (mtDNA), inherited only by the maternal lineage, showed that 62 percent of the population of Tacuarembó had Native origin, while 30 percent of the people of Cerro Largo and 20 percent in Montevideo had the same origin (Bonilla et al., 2004; Gascue et al., 2005; Sans et al., 2006). In addition, a sample of the entire country showed 31 percent maternal Indian ancestry (Pagano et al., 2005, modified). African maternal ancestry seems to be more homogeneous, being 17 percent in Tacuarembó, 21 percent in Cerro Largo, and 8 percent in the entire country (Bonilla et al., 2004, Pagano et al., 2005, Sans et al., 2006). No data for Montevideo has been published at present.

The studies about paternal lineages are scarce, but data have shown not more than 13 percent of Indian ancestry in Tacuarembó, and between 2 percent and 4 percent in Montevideo, depending on the markers used, single nucleotide polymorphisms (SNPs) or short tandem repeats (STRs) (Bertoni et al., 2005). A sample representing the entire country using STRs showed that only 1 percent of the population had African paternal ancestors and 5 percent, Indian (Pagano et al., 2005).

Unequal contributions of maternal and paternal ancestry, as well as higher values of Amerindian or African mtDNA related to autosomal polymorphisms-based estimations, reflect the same process that occurred in all Latin American countries mostly during colonial times: the favored unions were between Indian women and white men. Related to African contributions, the unequal contributions between sexes are not so clear, but again, unions between African women and European men were preferred (Sans, 2000; Sans et al., 2006).

Final Remarks

In Uruguay, perceived national identity is not reflected in current population genetic studies. Although some other factors might influence results, such as genetic drift or selection, generally genetic background is the result of the population history. A. Arnaiz-Villena (2001) states that the new technologies devised for studying gene variability in populations can result in more objective data than historical ones. Consequently, discrepancies between genetics and perceived national identity should be explained by distortions in the historical record. In this sense, A. M. Araújo Araujo (1994) has pointed out that identity is not based on objective elements, but on whom we want to be or, quoting Winston Churchill, "*History is written by the victors.*" Moreover, Uruguayan history has been written by the descendents of the European immigrants living in the capital city, primarily in the nineteenth century and affected by the Argentinean Domingo F. Sarmiento's ideas about barbarism and civilization; that is, promoting the extermination of Native populations and glorifying European culture.

Genetic data have shown that the contribution of African genes is around 6 percent (Hidalgo et al., 2005), a value that approximately coincides with the percentage of people that are self-ascribed as "black or mixed-black" "race or ethnic group" (Instituto Nacional de Estadísticas, 1997). However, these two results are derived from and link to very different realities. The first case refers to the percentage of genes in the populations, while the second refers to the percentage of people that self-ascribed as Afro-descendants. Genetic studies done in self-ascribed Afro-descendants showed that, in the northeast, the percentage of African genes is around 39–47 percent, while in the south, it is close to 47 percent (Sans et al., 2002; Da Luz et al., 2010). Accordingly, it might be expected that the 5.9 percent of the population who self-ascribed to that group contributed less than 3 percent (around 45 percent of the 5.9 percent) of the total African genes. However, the other part of the contribution (in this case, little more than 3 percent) should be explained by other facts, such as sampling errors, lack of knowledge of ancestors, or of African genes that came from other origins. In this sense, the determination of ancestry is limited by the fact that people do not know who their ancestors were, including often recent ancestors, sometimes as a result of adoption or as a result of other factors (Race, Ethnicity, and Genetics Working Group,

2005). Furthermore, African genes could come indirectly, for example from the Canary Islands, one of the most important origins for migration to Uruguay. This has been observed when analyzing the frequency of mtDNA haplogroup U6b1 (Sans et al., 2006). On the other hand, when analyzing the 2006 survey data about ancestry, 9.1 percent recognized that they had African ancestors, a value not far from the minimum determined by mtDNA (8 percent, modified from Pagano et al., 2005).

The analysis of Native gene contribution is more complex. In 1996–1997, only 0.4 percent of the population self-ascribed as Indian descendant, while in 2006, 4.5 percent indicated they have Native ancestors (Instituto Nacional de Estadística, 1997, 2006). Genetic data showed 10 percent of Native-Indian gene contributions (Hidalgo et al., 2005), while the maternal inherited contribution was around 31 percent (modified from Pagano et al., 2005).

Two aspects should be noted: the discrepancies between self-identification and genetics and the differences between the two surveys. Related to the first, it can be argued that Native ancestors lived during a distant time, and consequently could be easily ignored. This is also supported by the Indians' apparent invisibility, which together with the historical accounts that recounted their extinction has collaborated to inform perceptions of national identity. As E. Renan argued as early as in 1882, "*The essence of a nation is that all individuals have many things in common, and also that they have forgotten many things*" (1939: 191). Moreover, the fact that the main genetic contribution came from the maternal side (mtDNA), can be explained by the fact that the offspring of European or Creole men and Indian women rapidly integrated into the dominant part of the society, differently than African descendants, as happened in other parts of Latin America (Mörner, 1967). Regarding the second aspect of the differences between the two surveys, it is necessary to say first that, as the question changed from the first survey (self-ascription to a race) to in the second survey (ancestors from any "race or ethnic group"), it is difficult to compare both surveys. Moreover, differences can also be due to other factors, such as a change in identity or ethnicity over time.

In Uruguay, the revision of national history seems to be recent, as most of the studies were published after the 1980s. The antecedents can be seen in Eduardo Acosta y Lara's work, mainly published between the 1960s and 1980s. His books about the Charrúa wars (Acosta y Lara 1961, 1969) opened a new era in ethnohistory in Uruguay, and were first followed by González Rissotto and Rodríguez Varese (1982) and others. Simultaneously, genetic studies focused on non-European morphological markers such as Mongolian spot gave unexpected contributions by non-European populations (Sans et al., 1986, and subsequents). Moreover, some organizations of non-European descendants were founded in Uruguay in the last two decades: in 1988 Mundo Afro, the biggest organization that brings together African descendants was created in Montevideo, and one year after that, the Association of Charrúa Nation Descendants (ADENCH) was founded, which also includes descendants of other non-Charrúa Indians, and/or unknown Indians. Native-Indian descendants have begun to search for their origins in genealogical data, oral histories, and genetics, while other Native descendants have sustained *charruísmo*, defined by R. Pi Hugarte (2003) as the "uncritical exaltation" of the Charrúa Indians. However, related to changes between the two surveys, it is necessary to note that, when analyzing Afro-descendants, the percentage of African and their descendants increased slightly (5.9 percent to 9.1 percent), while Native descendants multiplied by 11 (0.4 percent to 4.5 percent).

As a conclusion, it should be emphasized that Uruguayan national identity is changing, and data coming from different disciplines are emerging in ways that collectively contribute to present a new more Latin American vision. This new vision will effect education (mainly related to history), medical practices (as inherited illnesses that originated in different parts of the world are investigated), self-ascription, and ethnicity.

Acknowledgments

The author is grateful to Pedro C. Hidalgo for his comments on an earlier draft of this paper. Special thanks are due to Susan Lobo for her valuable comments as well as for her assistance in the English version.

Note

1. The expected percentage is based on the frequency found in Southwestern European countries, as Portugal and Italy. The observed percentage was obtained in newborns from a hospital in Montevideo, Uruguay.

References

Acosta y Lara, E. (1961) *La Guerra de los Charrúas en la Banda Oriental (Período Hispánico)*. Montevideo: Monteverde y Cía.

———. (1969) *La Guerra de los Charrúas en la Banda Oriental (Período Patrio)*. Montevideo: Monteverde y Cía.

———. (1985) Salsipuedes 1831: Los lugares. *Revista Facultad de Humanidades y Ciencias (Montevideo), serie Ciencias Antropológicas*, 1, pp. 65–88.

Acosta y Lara, E. (1989) Salsipuedes 1831 (los protagonistas). *Revista del Instituto Histórico y Geográfico del Uruguay*, 24, pp. 73–104.

Aguiar, C. (1992) Cultura e Identidad: Una aproximación sociológica. In: Achugar, H. and Caetano, G. (eds.) *Identidad uruguaya: ¿mito, crisis o afirmación?* Montevideo: Trilce, pp. 167–171.

Araújo, A. M. (1994) *Montevideanos: Distancias visibles e invisibles*. Montevideo: Roca Viva.

Armelagos, G. J. (1995) Race, reason, and rationale. *Evolutionary Anthropology: Issues, News, and Reviews*, 4, pp. 103–109.

Arnaiz-Villena, A. (2001) Historic genomics: An emergent discipline. *Human Immunology*, 62, 869–870.

Azara, F. de (1809) *Voyages dans la Amérique Méridional*, 2. Paris: Dentu.

Barreto, I. (2007) Estudio biodemográfico de la población de Villa Soriano, Depto. de Soriano, Uruguay. Unpublished thesis (PhD), Universidad Nacional de Córdoba.

Bertoni, B. et al. (2005) Directional mating and a rapid male population expansion in a hybrid Uruguayan population. *American Journal of Human Biology*, 17, pp. 801–808.

Bilden, R. (1931) *Race Relations in Latin America with Special References to the Development of Indigenous Culture*. Charlottesville: University Press of Virginia.

Bonilla, C. et al. (2004) Substantial Native American ancestry in the population of Tacuarembó, Uruguay, detected using mitochondrial DNA polymorphisms. *American Journal of. Human Biology*, 16, pp. 289–297.

Cabrera, L. L. (1983) Los repartos indígenas de 1831. *Revista Antropológica (Montevideo)*, 2, pp. 31–34.

———. (1992) El Indígena y la conquista en la Cuenca de la Laguna Merín. In: Zubillaga, C. et al. (eds.) *Ediciones del Quinto Centenario*, 1. Montevideo: Universidad de la República, pp. 97–122.

Campagna, E. (1990) La población esclava en ciudades puertos del Río de la Plata: Montevideo y Buenos Aires. In: Nadalin, S. et al. (eds.) *História e População: Estudos sobre a América Latina*. São Paulo: Fundacão Sistema Estadual de Análise de Dados, pp. 218–225.

Carvalho Neto, P. (1965) *El Negro Uruguayo*. Quito: Editorial Universitaria.

Consens, M. (2003) *El pasado extraviado: Prehistoria y arqueología del Uruguay*. Montevideo: Linardi y Risso.

Da Luz, J. et al. (2010) Beta-globin gene cluster haplotypes in Afro-Uruguayans from two geographical regions (South and North). *American Journal of Human Biology*, 22, pp. 124–128.

Demasi, C. (1995) La dictadura militar: Un tema pendiente. In: Rico, A. (ed.) *Uruguay, cuentas pendientes*. Montevideo: Editorial Trilce, pp. 28–40.

Dobzhansky, T. (1955) *Evolution, Genetics and Man*. New York: Wiley.

El País Digital (2010) Nunca estuvimos tan unidos. *El País Digital*, July 13. Available from: http://www.elpais.com.uy/100713/ultmo-501589/ultimomomento/-Nunca -estuvimos-tan-unidos- [Accessed August 01, 2010]

Famighetti, R. (2000) *The World Almanac and Book of Facts 2000*. Mahwah: World Almanac Books.

Frega, A. (2008) Breve historia de los afrodescendientes en el Uruguay. In: Scuro Somma, L. (ed.) *Población afrodescendiente y desigualdades étnico-raciales en Uruguay*. Montevideo: Mastergraf, pp. 5–102.

Frega, A. et al. (2004) Esclavitud y abolición en el Río de la Plata en tiempos de revolución y república. In: UNESCO (ed.) *La ruta del esclavo en el Río de la Plata: Su historia y sus consecuencias*. Montevideo: Logos, pp. 115–148.

Gascue, C. et al. (2005) Frequencies of the four major Amerindian mtDNA haplogroups in the population of Montevideo, Uruguay. *Human Biology*, 77, pp. 873–878.

González Rissotto, R. and Rodríguez Varese, S. (1982) Contribución al estudio de la influencia guaraní en la formación de la sociedad uruguaya. *Revista Histórica del Museo Histórico Nacional (Montevideo)*, 54(55), pp. 199–316.

———. (1989) La importancia de las Misiones Jesuíticas en la formación de la sociedad uruguaya. *Revista Estudos Ibero-Americanos (PUCRS)*, 15, pp. 191–214.

———. (1994) Contribución al estudio de la influencia guaraní en la formación de la sociedad uruguaya. *Revista Histórica/Montevideo,*166, pp. 125–136.

Hidalgo, P. C. et al. (2005) Genetic admixture estimate in the Uruguayan population based on the loci LDLR, GYPA, HBGG, GC and D7S8. *International Journal of Human Genetics*, 5, pp. 217–222.

Huxley, J. and Haddon., A. C. (1936) *We Europeans: a Survey of Racial Problems*. New York: Harper.

Instituto Nacional de Estadística (1997) *Módulo raza: Período 1996–1997*. Available from: http://:www.ine.gub.uy/biblioteca/raza/MODULO_RAZA .pdf [Accessed May 12, 2009].

———. (2006) *Encuesta Nacional de Hogares ampliada 2006, Flash Temático 1: Ascendencia*. Available from: http://www.ine.gub.uy/enha2006/flash.asp [Accessed December 05, 2009]

Isola, E. (1975) *La esclavitud en el Uruguay desde sus comienzos hasta su extinción (1743–1852)*. Montevideo: Comisión Nacional de Homenaje del Sesquicentenario de los Hechos Históricos de 1825.

Long, J. and Kittles, R. A. (2003) Human genetic diversity and the nonexistence of biological races. *Human Biology*, 75, pp. 449–471.

Martínez Moreno, C. (1941) La esclavitud en el Uruguay. *Revista Nacional (Montevideo)*, 10, pp. 221–267.

Ministerio de Instrucción Pública (1925) *El Libro del Centenario del Uruguay: 1825–1925*. Montevideo: Capurro y Cía.

Montagu, A. (1942) *Man's Most Dangerous Myth: The Fallacy of Race*. New York: Columbia University Press.

Mörner, M. (1967) *Race Mixture in the History of Latin America*. Boston: Little, Brown and Co.

Padrón Favre, O. (1986) *Sangre Indígena en el Uruguay*. Montevideo: Comisión del Papel.

Pagano, S. et al. (2005) Assessment of HV1 and HV2 mtDNA variation for forensic purposes in an Uruguayan population sample. *Journal of Forensic Sciences,* 50, pp. 1239–1244.

Pi Hugarte, R. (2003) Sobre el charruísmo: La antropología en el sarao de las pseudociencias. *Anuario de Antropología Social y Cultural en el Uruguay 2002–2003,* pp. 113–121.

Pi Hugarte, R. and Vidart, D. (1969) *El legado de los inmigrantes,* 1. Nuestra Tierra, 29. Montevideo: Banda Oriental.

Pivel Devoto, J. E. (1959) Los orígenes de Paysandú. *Mundo Uruguayo,* 2073, p. 25.

Race, Ethnicity, and Genetics Working Group (2005) The use of racial, ethnic, and ancestral categories in human genetics research. *American Journal of Human Genetics,* 77, pp. 519–532.

Rama, C. (1967) *Los Afrouruguayos.* Montevideo: El Siglo Ilustrado.

Rela, W. (2000) *Uruguay: Cronología Histórica documentada, vol. 1, 1527–1810: Banda de los Charrúas / Colonización Española.* New York: Norman Ross.

Renan, E. (1882) (1939) What is a Nation? In: Zimmern, A. (ed.) *Modern Political Doctrines.* London: Oxford University Press, pp. 186–205.

Ribeiro, D. (1969) *Las Américas y la civilización.* Buenos Aires: Centro Editor de América Latina.

Rodríguez Kauth, A. (2009). Identidad social y nacional (América Latina: ¿mito o realidad?). *Topia,* September 22. Available from: http://www.topia.com.ar /articulos/identidad-social-y-nacional [Accessed April 07, 2010].

Rosenblat, A. (1954) *La población indígena y el mestizaje en América. I. La población indígena.* Buenos Aires: Nova.

Salzano, F. M. (1997) Human races: Myth, invention or reality? *Interciencia,* 22, pp. 212–226.

Sans, M. (2000) Admixture studies in Latin America: From the 20th to the 21st century. *Human Biology,* 72, pp. 155–177.

Sans, M. and Pollero, R. (1991) Proceso de integración uruguaya: El ejemplo de Tacuarembó. *Estudos Ibero-americanos (PUCRS),* 2, pp. 99–111.

Sans, M. et al. (1986) Presencia de mancha mongólica en recién nacidos de Montevideo. *Archivos de Pediatría del Uruguay,* 57, pp. 149–156.

Sans, M. et al. (1997) Historical genetics in Uruguay: Estimates of biological origins and their problems. *Human Biology,* 69, pp. 161–170.

Sans, M. et al. (2002) Unequal contributions of male and female gene pools from parental populations in the African descendants of the city of Melo, Uruguay. *American Journal of Physical Anthropology,* 118, pp. 33–44.

Sans, M. et al. (2006) Population structure and admixture in Cerro Largo, Uruguay, based on blood markers and mitochondrial DNA polymorphisms. *American Journal Human Biology,* 18, pp. 513–524.

Saura, A. (2008) *El Derecho Humano a la no Discriminación: Una aproximación a la situación de los y las afrodescendientes.* Montevideo: Ministerio de Educación y Cultura. Available from: http://www.mec.gub.uy/ddhh/pdf /ficha_3_08_interior.pdf [Accessed May 05, 1910].

Stalla, N. (2007) El largo drama desde la abolición hasta el reconocimiento como ciudadanos. In: Chagas, K. et al. (eds.) *Culturas Afrouruguayas*. Montevideo: Imprimex, pp. 4–8.

Steward, J. H. (1946) *Handbook of South American Indians,*1. Washington D.C: Smithsonian Institution.

United Nations (2007) Principles and Recommendations for Population and Housing Censuses, Revision 2. Statistical Papers, ST/ESA/STAT/SER.M/67/Rev.2. Available from: http://unstats.un.org/unsd/demographic/sources/census/docs/P&R_%20Rev2.pdf [Accessed June 04, 2010]

Verdesio, G. (1992) La República Arabe Unida, el maestro soviético y la identidad nacional. In: Achugar, H. and Caetano, G. (eds.), *Identidad uruguaya : ¿mito, crisis o afirmación?* Montevideo: Trilce, pp. 97–107.

VIñar, M. (1992) Memorias Fracturadas: Notas Sobre el Orígenes del Sentimiento de Nuestra Actual Identidad Nacional. In: Achugar, H. and Caetano, G. (eds.) *Identidad uruguaya : ¿mito, crisis o afirmación?l.* Montevideo: Trilce, pp. 33–47.

Wagley, C. (1971) The formation of the American people. In: Salzano, F. M. (ed.) *The Ongoing Evolution of Latin American Populations*. Springfield: C. Thomas, pp. 19–39.

World Cup News (2010) Posts Tagged "Garra charrúa": Uruguay's classy enforcer. May 16. Available from: http://www.worldcup2010world.com/world-cup-news/tag/garra-charrua/ [Accessed September 08, 2010]

Zubillaga, C. (1992) El aporte de la inmigración italiana en la conformación del movimiento sindical uruguayo. In: Devoto F. J. and Miguez E. J. (eds.) *Asociacionismo, trabajo e identidad étnica: Los italianos en América Latina en una perspectiva comparada*. Buenos Aires: CEMLA-CSER-IEHS, pp. 231–250.

10

Forced Disappearance and Suppression of Identity of Children in Argentina: Experiences in Genetic Identification

Victor B. Penchaszadeh

Introduction

This chapter deals with genetics and identity from a different angle than the others in this book. Indeed, it describes the use of forensic genetics in Argentina to help hundreds of children and young adults recover their identity after having been abducted as newborns or young children during the military dictatorship of 1976–1983. The fact that a book dealing with genetic admixture and notions of identity in Latin America dedicates a chapter to one of the most gruesome events that have happened during the twentieth century is a testimony to the dramatic history of the continent and to the role that socially committed geneticists have played in redressing crimes against humanity.

What will be described here is the saga of hundreds of children of *disappeared* and murdered political dissidents, who were appropriated and raised under a false identity by individuals linked to the military, sometimes by the very torturers and killers of their parents. The protagonists of this chronicle are the disappeared dissidents and their children, the relatives who never abandoned the search for their missing loved ones, the national and international human rights organizations that denounced the horror and carried out the research on the fate of the missing children, and the geneticists who implemented the methodology for the genetic identification of the missing children as they were found. The protagonist on the villain side is the military that reigned over life and death in Argentina and perpetrated abhorrent crimes against humanity, disappearing thousands of citizens and their children. The full understanding of the crucial role played by genetics after that period of madness, cruelty, and perversion requires a minimum of background.

Historical Background

Argentina, in the southern tip of South America, has a population of 40 million. At the beginning of the sixteenth century, the territory was inhabited by millions of Amerindians of at least 30 different ethnicities. During three centuries of Spanish conquest and colonization extensive admixture occurred, and the native population was further decimated in the late nineteenth century by an extermination campaign waged by the government. Amerindians currently number approximately 10 percent of the population. West Africans brought as slaves practically faded out through the nineteenth century, by admixture, attrition, and migration to neighboring countries, mainly Uruguay and Brazil. In the second half of the nineteenth century and first half of the twentieth century, Argentina received a huge influx of immigrants from virtually all countries of Europe, who intermixed with the existing population. In recent times, immigration has come primarily from neighboring countries with strong Amerindian ancestry (Paraguay and Bolivia). Reproduction patterns are largely open and admixture is extensive, although in smaller communities marriage within the same ethnicity is common. Currently, about 50 percent of the Argentine population trace their origin to Italian immigrants and 25–30 percent to Spaniards. In the city of Buenos Aires, the relative European, Amerindian, and West African genetic contributions to the gene pool have been estimated as 67.5 percent, 25.9 percent, and 6.5 percent, respectively (Martinez-Marignac 2004). Average consanguinity is less than 0.5 percent.

At the beginning of the twentieth century, Argentina's economy, based on agriculture exports, boomed, but social and political unrest were recurrent, due to perennial social injustice and exploitation of rural and industrial workers by the ruling class. The latter would resort to military governments and repression to control social unrest; between 1930 and 1973, the country suffered 30 military coups, with only one elected president ever being able to complete his mandate. Between 1966 and 1973, the military then in power increased repression, and when vast sectors of society became politically disenfranchised and economically deprived, armed opposition to the dictatorship by several radical groups began. The turmoil led to elections, and a civilian government was elected in 1973. However, it could not deal with the prevailing serious political and economical problems, and it turned to illegal repression to confront increasingly widespread social unrest. That is how killings of political dissidents by paramilitary squads directed by the government began, which in turn led to a growing radicalization from armed groups, and eventually another military coup on March 24, 1976.

The Military Dictatorship

The ensuing nine years of military rule (1976–1983) were characterized by a brutal dictatorship and state-sponsored terrorism directed against a wide spectrum of political dissidents. The targets of repression were primarily young workers, students, and intellectuals who were active in the struggle for democracy and social justice (CONADEP 1986).

The military carried out a systematic plan to exterminate all political dissidence "once and for all." Suspected activists, as well as their friends and relatives, were abducted violently from their homes in the middle of the night by military squads concealing their identity and under central command. Houses were ransacked of anything of value, and the victims confined to clandestine detention centers where they were subjected to torture and inhuman treatment. Thousands of abducted adults (the *disappeared)* were taken to one of several hundred secret detention centers run the military, where they were savagely tortured and, with few exceptions, eventually killed after variable periods of time in detention in abhorrent conditions. Most killed adults were buried clandestinely in unmarked graves or sedated and hurled alive from airplanes over the South Atlantic (Andersen 1993; Sims 1995). This genocidal extermination policy was never admitted or acknowledged by the dictatorship, who systematically denied the facts and its own responsibility, hypocritically attributing accusations of human rights violations to the action of "antipatriotic" forces (OAS 1978; Amnesty International 1979).

The figure of the *disappeared* was thus coined for the first time in the modern history of repression. Most of what was happening during the dark years of the dictatorship was unknown within the country, and only became public knowledge in Argentina after a civilian elected government took office in December 1983. By then, an estimated 30,000 people had been forcefully *disappeared* by the security forces.

A number of victims were living with their children when the security forces violently irrupted in their homes and abducted them. These babies and young children were abducted as well as "war booties." Furthermore, many abducted women were pregnant at the time of their disappearance, their number being estimated at about 500. These women were not spared torture and suffering during their captivity and delivered their babies in humiliating circumstances in clandestine detention centers or in military hospitals, only to be murdered shortly after delivery. It is estimated that no less than 500 babies were born in captivity to *disappeared* women (van Boven 1988; Anderson 1993; Abuelas 2008a; Harvey-Blankenship and Shigekane 2010).

The appropriation of babies of dissidents was part of a deliberate government policy, based on the military's conviction that "subversives breed subversives" and that they had the "duty of freeing these children from the subversive education" of their parents. The fate of these children varied according to the circumstances. Most commonly, these infants and young children were either kept as "war booty" by someone within the security forces, or handed over to childless couples associated with the repressive apparatus. Birth certificates were forged, and the children were registered as biological sons or daughters of their appropriators. Sometimes the complicity of a corrupt judiciary appointed by the military led to fast "legal" adoptions of supposedly "abandoned" children without checking whether they might actually be children of *disappeared* persons (Penchaszadeh 1992, 1997; Arditti 1999). At the very same time, however, relatives of disappeared people were filing inquiries in the courts about the fate of their loved ones, only to be falsely told that the government did not detain them and that nothing was known about them. In a few cases babies were handed over to neighbors with orders that they take care of them and threats of reprisal if they revealed their true identity. Occasionally, small babies were anonymously left in institutions for abandoned children. This gruesome reality was corroborated during the dictatorship by several international bodies like Amnesty International and the Organization of American States (OAS 1978; Amnesty International 1979).

The resistance to the dictatorship took many forms. Relatives of the disappeared, for instance, began to organize in search of their loved ones demanding that the government admits its responsibility for their fate and free them alive and unharmed. One of the first groups to surface publicly were mothers of disappeared persons who, in April of 1977, founded the organization *Mothers of Plaza de Mayo*. Soon the *Mothers* were joined by women who were searching not only for their sons and daughters, but also for their grandchildren. In October 1977, these women founded the

Association of Abuelas (grandmothers) of Plaza de Mayo (Penchaszadeh 1992; Arditti 1999; Abuelas 2008a) with the goal of finding their missing grandchildren. While a handful of them had known their grandchildren before their *disappearance*, in most cases the grandmothers only knew (or learned later through testimonies of survivors) that their daughter (or daughter-in-law) was pregnant at the time of her abduction by the military. The *Abuelas* began a detective work that would be their "trademark" through the years of the dictatorship and beyond. They gathered information on the missing children as reported by their relatives, such as name, sex, age, physical characteristics, pictures, date of disappearance, etcetera. In cases of disappeared women who were pregnant, they collected data from anonymous calls, reports from neighbors, visits to orphanages, suspicious adoptions, births alleged to have occurred at home, reviews of possibly forged birth certificates (such as those signed by physicians known to work for the military), and reports from witnesses of deliveries first occurred in military hospitals and detention centers. In each case, this data led to hypotheses on probable detention centers where the women were kept, the date of birth and sex of their babies, and information on appropriators. All this "intelligence" was painstakingly and tenaciously obtained and logged, with the conviction that when the rule of law returned, it would be an essential tool to find their grandchildren.

Sadly, the Abuelas would have to wait until the fall of the dictatorship before their demands of localization, identification, and restitution of their grandchildren would be heard. After several years of brutal repression and mismanagement of the economy, the military committed the last and fatal mistake of engaging in war with Great Britain for the possession of the Malvinas-Falkland Islands. Mounting civilian unrest after defeat of the military adventure eventually forced the regime to grant elections, won by a social democrat, Raúl Alfonsín, who was inaugurated as president in December 1983. One of the first acts of the newly elected government, early in 1984, was to appoint a National Commission on the Disappearance of Persons (CONADEP) to investigate the disappearances and produce data that would serve to take the former military rulers to court. The fate of the *disappeared* was in fact one of the major political and humanitarian issues facing Argentine society, as a number of mass graves were being uncovered and evidence kept mounting on the appropriation of babies. After months of work, the CONADEP concluded that none of the disappeared were alive, and its report laid the grounds for the prosecution of several top ranking military officers by a special Argentine tribunal, which found them guilty of gross violations of human rights, with convictions ranging from ten years to life in prison.

The Abuelas were invigorated by the political climate in which suddenly the question of human rights acquired such prominence. The abduction of children was repelled as actions against the most basic fabric of society, and most Argentines were united behind the quest of the Abuelas. Years later, the Argentine experience was cited in the reasoning behind articles 7 and 8 of the United Nations Convention on the Rights of Children, which recognized the right to identity as an essential human right of children. The challenges to redress the wrongs, however, were formidable. In the first place, the missing children had to be found, which was a very difficult task due to the network of deceit and complicity among the appropriators. Then, it had to be demonstrated beyond any doubt, that a particular child was indeed the offspring of a particular *disappeared* couple and, further, going through the legal procedures for the recovery of his/her true identity and obtain their restitution to the surviving biological relatives, most commonly grandparents, uncles, or aunts. The Abuelas themselves had done their homework, and in their travels obtained support from international bodies and from geneticists to implement genetic testing to prove genetic relationships. At the same time, relatives of the disappeared were demanding the identification of the remains of hundreds of people uncovered from unmarked graves. The hurdle, however, was that no forensic anthropology or genetic expertise existed in the country to identify the thousands of uncovered human remains or to identify the missing children.

The CONADEP asked then for technical assistance from the American Association for the Advancement of Science (AAAS), which in early 1984 sent to Argentina a group of forensic anthropologists led by the legendary Clyde Snow to work on the mass graves recently uncovered. Snow began to train a group of young Argentine anthropology students in the techniques of forensic anthropology and the first forensic identifications took place (with time, this group became the Argentine Forensic Anthropology Team: www.eaaf.org, who conducted a large number of forensic work associated with human rights violations, in Argentina and abroad) (Joyce and Stover 1991; Cohen-Salama 1992). At the same time, geneticists Cristian Orrego of the AAAS and Mary-Claire King from Berkeley, who were part of the AAAS delegation, assisted Argentine immunogeneticist Ana Di Lonardo, who ran at the time a histocompatibility testing lab in a city hospital, to set up genetic identification techniques to apply to appropriated children, as they were being found. At the time, the only markers available in Argentina to determine genetic relationships were HLA and red-blood group antigens. The statistical formulations commonly used in parentage testing were adapted to the fact that the parents of these children were *disappeared*, and in all probability dead, and that only putative grandparents were available for testing. The notion of *grandparentage index* was thus born, and six-year old Paula Logares, who had been appropriated when she was two-years old by a security officer linked to the military squad that disappeared and murdered her parents, was the first child of the disappeared to be identified by HLA testing and restituted to her biological grandparents (Di Lonardo et al. 1984). In the same year, the author of this chapter, an Argentine geneticist who was in exile in the United States, was commissioned by the Pan American Health Organization to go to Argentina and advise a recently created government commission on the technical and ethical guidelines to be followed in the genetic identification of appropriated children as they were found (Penchaszadeh, 1984). On the basis of his advice, the immunogenetic lab of the Durand city hospital started genetic testing (HLA and blood group typing) on several dozen potential grandparents of missing children, and developing a database with the results for future use.

It was in this new climate that the "intelligence" gathered by the Abuelas during years of detective work on the fate of hundreds of pregnant women in the concentration camps of the dictatorship and of the babies they had delivered, started to bear fruits. Circumstantial evidence (such as testimonies of camp survivors and of nurses that witnessed clandestine deliveries in military hospitals, analysis of suspicious birth certificates, reports of fake pregnancies and scrutiny of home births falsely alleged by appropriators, and much more) would point to the whereabouts of some missing children and served as the basis for suspicion and court action. Legislation was enacted to instruct the courts on how to process requests of relatives searching for missing children and, when circumstantial evidence provided a strong suspicion that a particular child could be the offspring of a *disappeared*, to order genetic testing of the child and its comparison with that of the suspected appropriators and the putative true relatives. The Abuelas assisted grandparents searching for their grandchildren with medico-psychological and legal teams and representation in the court (Abuelas 1995, 1997).

In May 1987, Congress enacted a law regulating all aspects of genetic testing for grandparentage testing on children presumed to be offspring of the *disappeared*. The law ordered the creation of a National Genetic Database in the Immunology Laboratory of the Hospital Durand of Buenos Aires, which would test and store genetic information on families that were searching for missing children and conduct genetic testing on the children as they were being found. Judges would order genetic testing on such families as well as on any child whose identity was in doubt or when there was insufficient evidence to suspect that a child could be the son or daughter of a *disappeared*. Most court cases were initiated by a demand submitted by the Abuelas on behalf of a particular family, when sufficient circumstantial evidence had been gathered on a particular child. The evidence was usually the result of research conducted by the Abuelas themselves and was based on historical information, physical resemblance, characteristics of the adults acting as parents (usually police or army officers), birth certificates signed by physicians known to work for the military, and other circumstantial data. It was then the judge's responsibility to review the evidence presented, to approach the couples acting as parents, and to order genetic testing to the relevant individuals, that is, the child, the couple acting as biological or adoptive parents, and the putative biological grandparents (Penchaszadeh 1997).

Developments in Genetic Identification

The establishment of family relationships by genetic testing is based in the analysis of the inheritance of particular genetic markers from putative parents to a putative child and are commonly used for paternity testing. The preferred markers for genetic identification in the early 1980s were the histocompatibility (HLA) antigens, produced by lymphocytes according to information encoded in various genes in the short arm of chromosome 6, the best known being the *HLA-A, HLA-B, HLA-C,* and *HLA-Dr.* The immense variability of the HLA genes makes it extremely unlikely that two unrelated individuals could share the same genetic combination.

Important developments in molecular genetics occurring in the 1980s and later transformed completely the field of human genetic identification. While a review of these developments is beyond the scope of this chapter, suffice it to state here that DNA variation is known to occur both in coding as well as noncoding segments of the genome, although the latter are much more common. DNA varies normally in the population in the length of some noncoding repeated segments, such as the *variable number of tandem repeats* (VNTRs or *minisatellites*) and the *short tandem repeats* (STRs or *microsatellites*), which are scattered throughout the genome and are extremely useful in genetic identification. Human DNA also varies, every about one thousand bases, in the sequence of nucleotide bases along its length, giving rise to *single nucleotide polymorphisms* (SNPs), of which there exist over 3 million in the genome. As nuclear autosomal DNA occurs in two copies (maternal and paternal), all persons carry two *alleles* (variant alternatives) at each site, which can be either identical or different from each other. Every parent transmits one allele of nuclear autosomal DNA to each of his/her child. Thus, each child carries one allele coming from the father and one from the mother, at every DNA site (Morling et al. 2002). In addition to the DNA markers in the *autosomal* (nonsex) chromosomes, there are markers in the Y chromosome that are used to determine the sex of the person from whom the sample comes, as well as to establish paternal lineage (Gill et al. 2001). Furthermore, DNA is also present in the *mitochondria*, cytoplasmic organelles inherited exclusively through the mother. Mitochondrial DNA (mtDNA) exists in single copy (*haploid*) and is identical along the maternal lineage, that is, in siblings and their mother, maternal grandmother, maternal aunts and uncles, cousins via the mother's sisters, and so on. Human mtDNA is highly variable in the population, particularly in its noncoding control region, where it contains a hypervariable segment. The chances that two unrelated individuals could have identical mtDNA sequence is extremely low, so its analysis is very useful for genetic identification through the maternal line (King 1991; Owens et al. 2002; Ginther et al. 1992).

The discovery and systematic study of human DNA variation came together with the development of an unprecedented ability to analyze genetic markers with ease and inexpensively, from minute quantities of DNA, and for many different purposes. Current forensic genetic analysis worldwide uses DNA testing as the gold standard and is completely standardized and automated. It is based in the analysis of STRs at a minimum of 13 loci along several autosomal chromosomes, plus sex chromosomes' STRs (Bar et al. 1997). Mitochondrial DNA analysis is used in special circumstances, such as when only maternal relatives are available, or to confirm nuclear DNA results, or in the analysis of remains, as mtDNA tends to be more stable (Carracedo et al. 2000).

With current molecular techniques, parentage or grandparentage testing can easily exclude with 100 percent certainty any nongenetic relationship. Conversely, if DNA analysis fails to exclude a genetic relationship, it will inform the probability that a particular individual is the child or grandchild of putative relatives (*probability of inclusion*), which is arrived at after complex mathematical calculations that take into account the relative frequency in the population of the particular alleles found at the analyzed loci. With modern DNA analysis of at least 13 single-locus or multiplex systems, inclusion probabilities of parentage or grandparentage typically reach 99.9 percent if the genetic relationship is true.

In Argentina, DNA testing for human genetic identification started to be implemented in the early 1990s, and the National Genetic Database eventually substituted DNA testing for HLA typing in all its forensic genetic testing. In the context of the missing children, the question is: Given circumstantial evidences suggestive that a particular child could be the son or daughter of a certain *disappeared* couple, can this be proven by testing putative grandparents? Since all genetic material passed by parents to their children derive from the grandparents, the genetic relationship between a child and putative grandparents could be determined with a very high degree of probability if all four grandparents are available, or if there are collateral informative relatives (Abuelas 2008b).

When the Missing Children Were Still Children

During the late 1980s and 1990s, when court proceedings requesting genetic identification of disappeared children began, the babies born in the late 1970s and early 1980s were still minors, and genetic testing and matching with putative grandparents' genetic markers were decided by the courts. A number of ethical considerations guided the procedures. In the first place, the well-being and the best interests of the children have always been the top priority. Teams of psychologists assisted the judges in taking the appropriate steps at the appropriate time to avoid further psychological trauma to the children (Abuelas 2008b). The premises were that these children were victims of crimes against humanity in the form of forced disappearance of their parents and abduction and suppression of identity, and that the state had the duty of bringing those responsible for these crimes to justice and most importantly to redress the harm caused to the child. Thus, it was sustained that these children had the right to recover their true identity, to learn the truth about their family history and the fate of their parents, and to live with their real family. These principles guided the actions of the Abuelas in the courts and in public in their quest to recover their grandchildren. In a few cases, missing children were found living with loving families who had adopted them in good faith, sometimes not even knowing that they were children of the disappeared. In these cases, adoptive and biological families usually reached agreements that would preserve the best interests of the child. In some instances, this meant that the child continued to live with the adoptive family, but having gained psychologically by learning his/her true identity and history and incorporating the biological family in his/her life (Abuelas 2008c).

   The location of a missing child, the court proceedings, the genetic testing, and the restitution processes were very complex, and many hurdles stood in the way, particularly when, as was usually the case, the child's appropriators were security officers or their accomplices. These individuals would thwart investigators and confront the inefficient Argentine judiciary, causing undue procedural delays, while keeping their grip on the appropriated child. The sloppiness of some judges even enabled in several instances the appropriators to flee the country with the abducted children. In the period ranging from the mid 1980s to the mid 1990s, dozens of trials were conducted to solve the identity of children found in suspicious circumstances in military families or their accomplices. DNA tests ordered by the courts revealed the true identity in about 50 children. Actions to preserve the best interests of the children thus identified included a strict observance of privacy of the proceedings, psychological support, and custody decisions that took account of each case's particular circumstances. In a few cases, the DNA of a child did not match with any of the putative grandparents whose DNA was stored in the National Bank of Genetic Data, leaving open the possibility that they could still be children of disappeared parents, whose relatives had not yet contributed DNA to the database.

These approximately 50 children identified as offspring of disappeared parents experienced suddenly a profound developmental and identity change of apocalyptic proportions that defied known concepts about mental health and plasticity of the mind. Their reactions and later development have been quite diverse, depending on a number of factors, some of which are not really known: circumstances of gestation and delivery, quantity and quality of time allowed to be with their mother before she was murdered, type of family in which they were raised, type of interactions between appropriators and child (there is some evidence of perverse relationships consistent with the notion of "war booty"). While personal details are largely private, it is known that some children had more psychological problems than others, and in general this was somehow correlated with age at identification (the older the child, the higher the chances of problems) and the adversarial position of their appropriators. Recorded experiences reveal that in most children, initial reactions of confusion and fear were soon followed by feelings of relief and liberation. Learning the truth about who their parents had been, about the fact that they had not been abandoned but stolen, and that they had loving relatives that had not ceased to search for them was a liberating experience for these children. The fact that most appropriators turned to lies in order to conceal their crime contributed to widen the affective distance between them and the children (Larmer 1993; Nash 1993).

In hindsight, 20 or so years later, most recuperated children have grown up as young adults without unusual problems in their mental health, happy to have recovered their true identity and history, and proud of who their parents had been and feeling affinity for the social causes they had espoused. Indeed, a number of them have turned social activists themselves, espousing social justice causes similar to those for which their parents lost their lives. Some are doing volunteer work at the Association of Abuelas, participating in their quest to find other children of the disappeared that were appropriated during the dictatorship. It would be disingenuous, though, to say that the horrendous experiences of these children and the struggles they went through in defining their identity left no marks. Several interviews given by these young men and women have recently been published (Argento 2008).

The Last Quarter of the Century

Along the 27 years of democracy and rule of law that have passed after the end of the dictatorship, many ups and downs occurred in Argentina in the area of human rights defense and reparations, as well as in the historical memory of the repression and in the quest for justice and against impunity. After the initial impulse of bringing to justice those military responsible for the systematic extermination campaign and of successful trial and conviction of the nine top generals of the military juntas (former dictator Videla was convicted to life in prison), Alfonsin's government gave in to pressure from the military establishment and led Congress to vote the laws of "full stop" and "due obedience," which suspended more than 600 ongoing trials of officers accused of disappearances, torture, and murder, and in effect meaning the prescription of the crimes against humanity (the only crimes that were exempted from impunity were those related to the appropriation of children). Alfonsín's successor, Carlos Menem, was even less keen in confronting the military nor interested in human rights policies and pardoned all the generals that had been convicted only a few years earlier. The only trials that could take place were those related to appropriation of children, which was taken advantage of in 1996 by the Abuelas, who initiated a complaint as plaintiffs, arguing that the stealing of babies was actually a criminal conspiracy directed by the highest echelons of power to steal the children of dissidents. After 11 years of proceedings, the courts sided with the Abuelas and sent several high-ranking military officers to prison, including former dictator Videla, who returned to jail for life.

In order to speedup court proceedings for the identification and restitution of disappeared children and to ensure the observance of articles 7 and 8 of the International Convention on the Rights of the Child (pertaining to the right to identity), the government created in 1992 a National Commission for the Right to Identity (*Comisión Nacional por el Derecho a la Identidad*, CONADI). This commission began functioning within the Ministry of Justice and Human Rights, with delegations in all provinces, and was given the task of dealing with all aspects of the defense of the right to identity of children and adults. In addition to processing individual inquiries about identity of children of the *disappeared* during the dictatorship, it deals with traffic of children and the stealing and selling of babies, which are unfortunately a widespread problem in Argentina. In accordance to its mission, CONADI receives requests from adults who have doubts about their identity because they were born during the dictatorship and learned that they were in fact adopted, or have other reasons to suspect that they could be children of the *disappeared*. CONADI works in close connection with the courts and the National Genetic Database, where it refers individuals for genetic testing and comparison of their DNA profiles against those in the database from relatives of the disappeared, to detect possible matches.[1]

Important legal breakthroughs, occurred in Argentina beginning in 2003, related to human rights violations and impunity. Nestor Kirchner became president in 2003, and started a proactive policy of uncovering the egregious violations of human rights during the dictatorship and taking steps to stop the impunity that had prevailed until then for such crimes. It should be noted here that, according to the Rome Statute of the International Criminal Court, which came into force on July 1, 2002, when committed as part of a widespread or systematic attack directed at any civilian population, a "forced disappearance" qualifies as a crime against humanity, and thus is not subject to a statute of limitations (International Criminal Court 2002). On December 20, 2006, the United Nations General Assembly adopted the International Convention for the Protection of All Persons from Enforced Disappearence (United Nations 2007). President Kirchner's government ratified the Convention on the Non-Applicability of Statutory Limitations to War Crimes and Crimes Against Humanity, and in 2005, Congress made the crucial decision to repeal the two impunity laws (of "full stop" and of "due obedience") that had been in place for 20 years preventing the prosecution of hundreds of responsible for crimes against humanity. Further, shortly thereafter, the Supreme Court declared such laws to be unconstitutional enabling the removal of the pardons that former president Menem had granted to the generals convicted in 1985. This allowed the reinitiation of hundreds of prosecutions to former military officers accused or torture, disappearance, and murder. At the time of this writing, 649 former military officers or security agents are being prosecuted, of whom 421 are detained and 228 in freedom. Two hundred thirty accused died while they were being investigated and 18 were declared incompetent, including former dictator Massera. So far 23 trials have concluded with 68 convictions and 7 acquittals, and 10 additional trials are in process involving 59 military officers and security agents. While some decry the slowness of this process ("justice delayed, justice denied"), the incredible importance of these trials cannot be overemphasized. It gives society a chance for reconciliation with the judiciary, some degree of healing and reparations to victims and their relatives, and is a very important step toward the establishment of a collective historical memory, without which there is always the peril of new fracturing of the establishment of democracy. A special trial on the appropriation of children has started in March 2011 where the military's top-ranking officials are again standing trial.

When the Disappeared Children Turned Adults

As time passed, those abducted babies became young competent adults that could decide for themselves how to proceed if they had suspicions about their origin or doubted their identity. Similar to the trials against those who ordered or committed crimes against humanity (including the appropriation of children) the courts would receive claims of disputed identity by a growing number of young adults who suspected they could be children of the disappeared and wished to allay their doubts and eventually find their true relatives. Indeed, a prediction 20 years earlier by Estela Carlotto, president of the Abuelas, was being fulfilled *("now we are searching for them, in the future they will search for us")*.

The circumstances through which dozens of offspring of the disappeared learned their true origin and identity as adults have been diverse. In part, it was led by the natural drive for identity searching that adoptees show worldwide, whether or not their adoption status had been disclosed. In the Argentine context, however, the phenomenon of the appropriation of children of the disappeared by the military or their accomplices is so much in the public domain, suspecting or knowing that one was adopted during those years immediately raises the question: could I be a child of disappeared parents? Further, in this social climate it is very difficult for appropriators to conceal such gruesome secrets for too long. Sooner or later, a misstep will be committed or a witness may come forth, triggering circumstances that led many young men and women to doubt their identity. Most commonly they sensed that "something was wrong" in the household, such as domestic violence, mistreatment, deceit or lies in the answers to questions about their origins, or simply lack of physical resemblance. Anonymous reports by witnesses, after decades of silence, played a role in a number of cases (such as from nurses that had witnessed clandestine deliveries or from neighbors that saw a baby being brought to a home, or simply from relatives or acquaintances that decided to come forward). Public campaigns by the Abuelas ("If you were born during the dictatorship and doubt your identity, call us!") undoubtedly awakened the conscience of many. The following illustrative cases highlight the complexities of human development and identity in these circumstances.

Juan Cabandié-Alfonsín was born in March 1978 to her 17-year-old disappeared mother Alicia Alfonsín, in the infamous Navy School (ESMA, for its acronym in Spanish). According to survivors from that illegal detention center, she named her son Juan. Juan's father, Demián Cabandié, aged 19, was also disappeared by the military shortly afterward and never seen again. At 15 days of life, Juan was taken off his mother, who subsequently was murdered and never seen again, and stolen by Luis Falco, a military officer who raised him as a biological son, Mariano Falco, concealing his true origin. As a child, Mariano/Juan befriended twin boys of similar age who supposedly were the children of another military officer, Samuel Miara. The case of these twins became public news between 1984 and the mid 1990s because they too were appropriated children, and the court proceedings leading to their identification and restitution a number of years later were polemic and captured the attention of society and media. So Juan first learned watching TV as a youngster, that the best friend of his "father" was a confessed appropriator of the two children who were his friends (Samuel Miara is now in prison for torture). This serendipity, coupled with a history of mistreatment and disaffect on the part of Mr. Fialco, who boasted his participation in house raids and kidnappings of "subversives," helped Juan to tie pieces together and fueled his suspicion of not being the biological child of Mr. Falco and his wife. Juan was 25 when in 2004 he asked his "mother" whether he was adopted, and when she admitted it, he severed his ties with the appropriators and went directly to the office of the Abuelas to request DNA analysis. His grandparents' DNA profiles were in storage at the National Bank of Genetic Database and soon thereafter Juan Cabandié-Alfonsín recovered his identity, his real name, and his history. He is now a progressive political activist and an elected legislator of the city of Buenos Aires.

Another emblematic case is that of *María Eugenia Barragán-Sampallo*, daughter of Mirta Barragán and Leonardo Sampallo, both leftist activists. Mirta and Leonardo were kidnapped by an army squad in 1977, never to be seen again. They had a three-year-old son, who was left in a police precinct as his parents were kidnapped (he was soon picked up by relatives). In addition, Mirta was six months pregnant. She was kept in an illegal detention center and in March-April 1978 gave birth to María Eugenia, probably in the Military Hospital. The baby was handed over at three months of age to the couple Rivas-Gomez by a military officer who was a friend of the woman and was registered as a biological daughter of the couple using a fake birth certificate. By her own account, María Eugenia learned at age seven that she was not the biological child of the couple, but her questions about her real parents and her origin were always answered with fabrications and lies. Through an anonymous report the Abuelas found the child, but genetic testing ordered by the courts at the time did not match with anyone in the database of the National Bank of Genetic Data, because no samples of the paternal side were stored (it was still the pre-DNA era). After years of alleged mistreatment, María Eugenia left the home of the couple as a teenager, severing her ties with them. She eventually recovered her identity in 2001, when she volunteered to be retested, learning then that her maternal grandmother and her brother Gustavo had been searching for her for 24 years. Maria Eugenia's case is conspicuous in that she is the first child of the disappeared who sues her appropriators in court. Indeed, after a trial that lasted several years, on April 4, 2008, the Oral Tribunal No. 5 convicted the appropriators of illegal retention and hiding of a minor and of forging public documents to conceal the girl's true identity and sentenced them to eight and seven years in prison, respectively. The army captain who had handed the baby to them, Enrique Berthier, was convicted of kidnapping and sentenced to ten years in prison. In a press conference after the sentencing, María Eugenia differentiated adoption from appropriation with the following words:

> Can someone who stole a newborn, who concealed from her that she had been stolen, who might have disappeared or tortured her parents, who kept her away from them and her family, who always lied to her regarding her origin, who frequently mistreated, humiliated and deceived her, really feel parental love? My answer is no, that this type of relationship is marked by cruelty and perversion and not by love (Vales 2008; Elkin 2008).

Francisco Madariaga-Quintela is the son of Abel Madariaga and Silvia Quintela, who were political activists during the dictatorship. Silvia, a physician, was pregnant when she was abducted and *disappeared* by the military in January 1977 while Abel was able to escape to Brazil. Silvia gave birth in a military barrack to a baby boy in July 1977 and was never seen again. The baby was handed to a military officer who raised him as his own child with other biological children of his and his wife. When Abel returned to Argentina after many years in exile, he started to search for his son and became the executive director of the Association of Abuelas of Plaza de Mayo. Francisco grew up in a house where violence was an everyday experience, and where he always felt mistreated and outcast. He began suspecting that he could be a child of the disappeared when the couple admitted that he was not their biological son. In February 2009, at 32 years of age, he went to CONADI stating that he doubted his identity and requested DNA testing and comparison with the existing database of relatives looking for disappeared children. Francisco's DNA was tested at the National Bank of Genetic Data and a match was found with Abel Madariaga. The reunion of Abel, who found the son he had been searching for three decades, and Francisco, who met his biological father after suspecting for years that he was not the son of the repressor who had raised him, was incredibly emotional. This author, witnessing the reunion between father and son, and interviewing Franciso, could only see joy and elation in him. ("It is so wonderful to have your own identity," "I have been born again," he declared to the media) (Vales 2010; Martinez 2010).

While increasing numbers of young adults who doubted their identify were coming forward requesting DNA testing to solve their doubts, the Abuelas continued to investigate the whereabouts of the missing children of the disappeared, and when they had enough circumstantial evidence that someone could be the offspring of a *disappeared*, they initiated complaints in the courts. Given that the possible victims were already adults, genetic identity testing could not be ordered without their consent, which, not surprisingly, was not always immediately forthcoming. Indeed, many of these young adults were living relatively normal lives and their initial reaction to the abrupt news that they could be children of the disappeared was received with a mix of anguish and disbelief. Undoubtedly, they had to confront two awful realizations: that the parents they have known had raised them illegitimately, and that, in some cases, their adoptive parents could have participated in the deaths of their biological parents. These realizations had different consequences, depending on a number of very individual circumstances, including the relationship with their "rearing parents," how they had been treated, whether or not they had been told that they were not biological children, or whether or not they had been lied regarding their true origin and identity. Some young men and women resorted to denial and to increased attachment with their appropriators. On the other hand, others met the possibility of being a child of a disappeared as a challenge that had to be dealt with immediately, sensing something amiss in the relationship with those who had reared them. In most cases, however, initial responses of disbelief, denial, and resistance to genetic testing were slowly, albeit with ups and downs, followed by consent to testing.

The position of the judiciary regarding the need for consent from competent adults for genetic identification testing became more nuanced with time. In 2003, *Evelin Vasquez-Ferrá*, a young woman suspected by the Abuelas as the daughter of disappeared parents being raised by a navy officer, declined to volunteer a DNA sample to prove the case (Argento 2008). The Supreme Court sided with the woman on the grounds of her right to privacy and that she could not be forced to provide incriminating evidence (her DNA) that could used to convict those whom she considered her parents. This ruling was criticized by many who felt that that the justice system has a special duty of investigating and establishing the historical truth, given that forced disappearance is a crime against humanity and that in these cases the responsibility of the state to solve such crimes should overrule the rights of a particular individual. The psychological and ethical conundrum here was that the subject of those individual rights who resisted testing was herself a victim of a crime against humanity. Later the court, with a different composition, allowed that when suspected offspring of appropriators declined to provide voluntarily a DNA sample for genetic identification, judges could obtain DNA samples by non invasive methods, such as from personal belongings (tooth brushes, underwear, etc). Through this procedure, Evelyn Vasquez-Ferrá was eventually identified as the daughter of Rubén Bauer and Susana Pegoraro who had been disappeared by the military in 1977 (Argento 2008). This case is probably the most extreme so far of a child of the disappeared who, while expressing solidarity with the grandparents who were searching for her, and even willingness to research her genetic identity, resisted doing so voluntarily to avoid incriminating those whom she continues to consider as her parents.

Victoria Donda-Perez was born in 1977 in the infamous ESMA, the clandestine detention center of the navy, to María Perez, an activist in the Montoneros movement, who was pregnant when she was disappeared. Her father, José Donda, was also disappeared by the military shortly afterward. According to the report of survivors, María named her baby Victoria before being murdered. The baby was appropriated by a torturer at the navy center, José Azic, who named her Analía. Analía/Victoria grew up believing she was a biological daughter of this couple, and that she was a biological sister to another, unrelated appropriated girl. What makes this case notable is that in her youth Analía/Victoria became a leftist activist, to the dismay of her appropriator. It was in these activities that her fellow activists began suspecting that she could be a daughter of the disappeared because her appropriator was known as a torturer during the dictatorship. Eventually in July 2003, Mr. Azic was named by Spanish judge Garzón in an extradition of 24 officers accused of torture and other crimes against humanity to be judged in Spain. Analía/Victoria learned the truth about who Azic was when he shot himself the day the list was made public, albeit failing in the suicide attempt. From then on, the life of Analía/Victoria was turned upside down, torn between the love for the couple who had reared her and the admiration and identification with the ideas and the courage of her disappeared parents. In her autobiography (Donda 2009), she uses the allegory of *"as if my mother's genes were acting in me"* to try to explain why, even before knowing her true identity, her personality, political convictions, and social activism were so similar to her mother's. As she tells it, "Analía turned into Victoria, without letting Analía go," and it took her a full year after learning that she was not who she had been told she was, before deciding to undergo DNA tests, which confirmed that she was indeed Victoria Donda-Perez. This case has a number of additional facets that epitomize the tragic recent Argentine history, the plight of the disappeared, and the terrible sequelae in the fabric of society. Since for reasons of space I cannot go into details, suffice it to say here that (a) the brother of Victoria's father was a torturer at ESMA and was probably involved in the detention of María and José and in handing over Victoria to Azic (he is now in prison while his trial is in process); (b) a previous daughter of María and José, two years older than Victoria, was adopted legally by the torturer uncle and has had opposite attitude to that of Victoria toward her biological parents; (c) Victoria continues being haunted by emotions derived from her history, and she publicly states that, while she considers that Mr. Azic committed crimes against humanity for which he should be punished, she still loves him; (d) Victoria continued her leftist political activism and was recently elected national congresswoman and as such she presides it's commission of Human Rights (Donda 2009).

Liliana Fontana was 20 when, several months pregnant, was kidnapped with her companion, Pedro Sandoval, and thrown into a torture center, and then was later transferred to a military facility to give birth. Right after delivery Liliana and Pedro vanished and were never seen again. The baby was appropriated by a navy intelligence officer, Victor Rie, and reared as their biological son. A number of circumstances directed the Abuelas' attention to this couple, including overhearing once that the supposed mother admitted to a friend that Alejandro was adopted. During the trial against the appropriation of babies pushed by the Abuelas in 2002, Victor Rei was summoned to give testimony. Two years later, realizing that the political climate had changed and fearing conviction, he decided to tell Alejandro that he was indeed born in a military barrack, and that he was "saved from death" by a fellow officer who brought him home. Alejandro, then 26, thought he had received real love and should not do anything that could incriminate his appropriators, so he did not cooperate when he was asked for a blood sample to compare his DNA with that of the grandparents of the disappeared. The judge, however, conducted a raid in his home and obtained personal objects from which DNA profiles were obtained and matched with those stored at the National Bank of Genetic Data, which proved that Alejandro was the son of Juliana Sandoval and Pedro Fontana. At the trial of Victor Rei, Alejandro started to drift emotionally toward his biological family (who by then he had already met), in part because his appropriators continued to be disingenuous. He began to process and incorporate what had happened to his parents, and "when you begin to realize that the person who raised you was a participant in that situation, you feel everything: pain, anger, sadness, rage" (Sandoval-Fontana 2010). Victor Rei was convicted in 2009 for appropriation and suppression of identity of a minor and sentenced to 16 years in prison. Although his wife was not tried, Alejandro severed his ties with her and drew himself closer to his grandparents. He volunteered a blood sample to be retested (which confirmed his identity) and is dedicated to piecing together the past and coming to terms with what happened to him and who he is. "Today, in spite of the tortures suffered in the concentration camp with my mother and father, I am a happy guy, I am myself, with the old history, the new history and the history that will come, I fight for the same ideas that my parents fought, because that was passed to me through their DNA."

Probably the most conspicuous current case, full of political facets, is that of *Felipe and Marcela Noble*, who in 1977 were adopted as infants legally, albeit without complete transparency, by Mrs. Herrera de Noble, one of the wealthiest persons of Argentina, owner of the multimedia Clarin, and currently a major opposition force in the country. These two young individuals (who are not related to each other) have been suspected for long of being children of disappeared parents but have been resisting for years in the courts to allow their DNA to be tested and compared with that of the grandparents of disappeared persons stored in the National Bank of Genetic Data. After an unsuccessful home raid in which DNA recovered from underwear was contaminated, the intervening judge treated the individuals as possible victims of being children of disappeared parents and ordered that they must provide, willingly or unwillingly, biological samples for DNA analysis and matching with the National Genetic Database of relatives of the disappeared. This case is further complicated politically and economically, as the adoptive mother claims "political persecution" without evidence, and Felipe and Marcela are heirs to her multimillion fortune. Mrs. Noble could end up in jail if it is proven that Marcela and Felipe are indeed children of disappeared parents, while the latter could loose their inheritance rights if the adoption is nullified. In June 2011, after several legal setbacks Felipe and Marcela agreed to provide DNA samples for comparison with the profiles stored at the National Genetic Database, where no match has been found with relatives of disappeared children up to the time of this writing. This case epitomizes the complex political and ethical nuances that genetic identification can have in Argentina.

The National Bank of Genetic Data

The Abuelas of Plaza de Mayo have been always at the forefront of the struggle to have governments take responsibility for the disappearances committed by the military, prosecute and convict the criminals, and take proactive measures to redress the wrongs. The plight of the Abuelas has been supported by the overwhelming majority of Argentine society, and thus the search for and genetic identification of the disappeared children has been a matter of public policy since the recovery of democracy. The National Bank of Genetic Data (*Banco Nacional de Datos Genéticos*) evolved from a small city hospital immunogenetics laboratory into a full-fledged forensic genetics center along its 25 years of existence. It was created by law of Congress in 1987, to deal with the identification of victims of human rights violations, and specifically to identify children of disappeared parents. By law, the DNA testing of all cases of disappeared children ordered by the courts must be performed at the bank's lab. In addition, the bank stores a database of DNA samples and profiles of putative grandparents and other presumed relatives of the disappeared, which is confronted when looking for a match with the DNA profile of every case in whom genetic identification is requested. Currently, the database consists of DNA samples and profiles of members of 311 families of the disappeared in which there is high suspicion or evidence that the disappeared woman was pregnant at the time of her abduction. So far 105 individuals have recovered their true genetic identity, 54 when they were still children and 51 already as adults. DNA samples from many more additional individuals did not find a match with the database, meaning that either they are not children of the disappeared or that the database does not contain all the families of the disappeared children. Given the estimation that the total number of disappeared children approaches 500, it is clear that the database is incomplete. Since the creation of CONADI, many identity requests coming voluntarily from young men and women are referred directly to the bank, without intervention of the courts. Only when a match is found, the case is referred to the judiciary. The bank has received and processed so far 2,500 samples from individuals referred by CONADI, most turning to be negative matches, with some notable exceptions.

Since its inception in 1987, and until recently, the National Bank of Genetic Data was under the joint jurisdiction of the National Ministry of Health and the city of Buenos Aires. In November 2009 a law of Congress approved a law transferring it to the jurisdiction of the Ministry of Science and Technology, a process that is being implemented at the time of this writing.[2]

Influence of Genetic Identification of Disappeared Children on the Development of Human Genetics in Argentina

Modern genetics in Argentina was developed initially in the 1960s in the School of Sciences of the University of Buenos Aires, mainly by cytogeneticists and Drosophila geneticists trained in the United States. In the 1970s, a few physicians trained abroad in the incipient discipline of medical genetics and set up clinical genetics and cytogenetics units in some hospitals. During the dictatorship, much of that effort went astray as most of those geneticists were forced into exile. When democracy was reinstituted, the human genetic capacity in the country was reduced to a few small groups in academic centers and hospitals. The creation of the National Bank of Genetic Data to address the tragedy of the disappeared children led to a surge in interest in the science of genetics by young biologists and physicians. This coincided with major scientific developments in molecular and clinical genetics internationally, which fostered a flurry of activity in basic and medical genetics in academic centers and medical care institutions in Argentina. Furthermore, forensic genetics and forensic anthropology became a natural magnet for young scientists, who added the novel genetic technologies in the identification of remains of the disappeared to classic forensic anthropology methods (EAAF Annual Reports 1990–2009; Joyce and Stover 1991; Cohen-Salama 1992). At the same time, the use of genetics to solve issues related to violation of human rights contributed to legitimize a discipline that itself had a past of human rights violations abroad in the forms of eugenics, genetic discrimination, racism, and "racial hygiene." For these reasons, genetics in Argentina became popular in society, which made it part of the public discourse in approving terms. This, in turn, fostered the development of research and applications in population genetics, forensic genetics, and human genetics in general, particularly its medical applications, that is, diagnosis of genetic disorders and genetic counseling (Penchaszadeh 2008).

Legal, Ethical, and Psychological Issues

The horrors of military dictatorship have left deep wounds in Argentine society at large. In particular, the treatment of children as war booties caused tremendous psychological and social harm to the children them-selves and their surviving relatives. The goal of genetic identification was originally thought as the most difficult hurdle for the recovery of miss-ing children. It turned out, however, that this was actually the least of the problems to solve. More important were the legal obstacles derived from a corrupt and inept judiciary and the political pressures from the government of President Menem (1989–1999) who did not welcome a public discussion of the human rights violations of the former rulers and its consequences. It was only in 1992, after multiple pressures from the Association of Abuelas and public opinion, that the government appointed CONADI with the task to promote the search of the missing children.

The types of challenges faced by searching families, governments, and the judiciary when the children were still minors were somewhat differ-ent from the ones presenting when those children were already compe-tent adults. The main ethical issues in the search and identification of missing children were to define the legitimate interests of the surviving relatives and the best interests of these children. There is no doubt that the former had a legitimate interest in recovering a missing child who was made an orphan by the criminal acts of the military and who was suffer-ing from a prolonged kidnapping by appropriators. From the angle of the child, it was felt that his/her best long-term interests were to recover his/her identity and personal and familial history, and to cease to live in a perverse environment of violence, lies, and deception. When disappeared children were still minors, actions were taken to minimize any possible short-term psychological harms of revealing the true identity and resti-tuting the child to the biological relatives. The Abuelas were among the first to recognize that extreme caution had to be exerted before decisions were made about the identity, custody, and living arrangements of these children. Indeed, when adoptions had been in good faith, while children learned their true identity and benefited from establishing loving rela-tionships with their surviving relatives, they usually remained with their adoptive families.

From the beginning, human rights groups and concerned lawyers, psychologists, and geneticists differentiated appropriation (a criminal action) from adoption. Indeed, the appropriation of children was com-mitted violently against the will of the parents, who not only had never abandoned their children, but also were ultimately murdered. Moreover, the surviving relatives who searched indefatigably for the missing chil-dren were systematically denied their parental rights. Further, the abduc-tions of children were not isolated events but part of a systematic plan that included assassinations, tortures, forced disappearances of adults, ransacking of their homes, and so on. The ideological justification of the abduction of minors was admitted by former army officers, who claimed that since "subversives educate their children for subversion" it was the army's duty to find "better families" for them (Penchaszadeh 1997). Since in many cases the appropriators had direct or indirect links to the murderers of their parents, the lives of these children were a prolongation of a state of captivity and disappearance. Finally, the perverse nature of the appropriator-child relationship, based in concealment of the truth and the constant, albeit hidden, reminders of the violent and illegal origin of the appropriation, perpetuated the initial psychological trauma, leading to a number of developmental problems detected in recovered children (Abuelas 2008c).

The process of restitution to the biological family was always preceded by psychological assessments of the child and his/her circumstances and measures were taken to minimize trauma. Through the law and the true love of the legitimate family, and proper psychological and social supports, restituted children were able to recover their identity and family history and developed healthy relationships with their true relatives, beginning a new and empowering life based in truth and justice (Abuelas 2008c).

The finding of disappeared children when they are already adults presents a different set of ethical and legal challenges, since as adults they can exercise their autonomy in decision making. A key ethical and social dilemma is that of truth versus the right to privacy. As mentioned above, many young men and women did not have this dilemma, for them solving the doubt of their genetic identity was a priority. Several others, though, who were happy with the life they were living, felt that the plight of their probable blood relatives was an intromission in their lives. After a number of legal turns, in late 2009, the Supreme Court has settled the dilemma between truth and personal privacy, dictating that blood sampling for DNA testing affects personal privacy and cannot be obtained in competent adults without consent (except from accused criminals when their DNA is part of the evidence in trial). However, it did allow the obtention of DNA by "non invasive" methods, such as testing personal objects like toothbrushes or underwear, without consent to prove identity of disappeared children. Shortly afterward, this ruling was made a law by Congress. What is at stake in Argentina is: (a) whether the right to privacy of an individual can trump the right of surviving relatives to learn the fate of their children and grandchildren, and (b) whether a fundamentalist approach to individual rights can prevent society to solve crimes against humanity, prosecute and convict those responsible, and redress the crime of kidnapping and suppression of identity of children.

The experience so far has been that the overwhelming majority of young men and women confronted with this dilemma opted for learning the truth. In most of the few instances in which there was initial resistance, the individuals came to terms with their own need to learn the truth and ended up volunteering for DNA testing. It is interesting, though, that a sizable proportion of these young adults have maintained relations with the families that reared them, particularly with the "mother." Psychological problems no doubt exist, but they have not been studied systematically.

Genetics and Environment in the Shaping of Personal Identity

It is impossible to describe in depth the horrifying experiences of the 105 captive children who, at varying ages, learned that they were not the children of those whom they had considered until then their parents, that their true parents had been murdered by the dictatorship, that the appropriators that raised them were in many instances directly or indirectly responsible for the disappearance and murder of their parents, that they were lied to consistently about their origin and ancestry, and concealed the fact that they had biological relatives who had been searching for them for decades. While the experiences of these children and young adults have been shocking, terrible, and dreadful, learning the truth has been for most of them liberating, healing, and energizing. Each case, however, is unique and distinct from all others, depending on a number of circumstances: who their parents were; why and how they were abducted, disappeared, and killed; where were they born; who appropriated them, how were they raised, and what they were told about their identity; how old were they when they learned their true origins; the nature of the process by which they recovered their true identity (ranging from a legally enforced testing of a child against the opposition of the appropriators to a voluntary search by a young adult suspecting of his/her true ancestry).

A key factor in the reactions was the age of the child at the time of disclosure and whether there existed loving and suitable biological relatives (usually grandparents) to take on the custody and rearing of the child. In general, the younger the child the better has been the passage from a usually perverse rearing environment in the hands of appropriators to a loving and supporting relationship with close relatives. Most of those identified during childhood accepted their true identity with relief and no evidence of trauma attributed to identification. Young adults, on the other hand, had more diverse reactions, ranging from initiating the requests for identity testing to resisting it all the way to the Supreme

Court. The search, localization, identification, and restitution of the true identity of the children of the disappeared is based on the notion that the forced disappearance of children is a crime against humanity and that the state has the duty to protect its citizens against these crimes, investigate them, prosecute the criminals, and bring justice to the victims. The experience has shown that knowledge of the truth, painful as it was, was a liberating experience for most these individuals. The joy of learning their true identity, however, has not prevented some of these young individuals from maintaining some links with their former appropriators, and even to express feelings of love for them (Argento 2008), attesting to the complexities of human nature and the reactions to extreme adversity. On the other end of the spectrum, most of the young men and women who recovered their identity, completely severed their ties with those who for many years they had considered their parents, to the extreme of sueing them for their appropriation or testifying against them in their prosecution. In any case, accepting their true identity was for most children and young adults a long process, full of ambiguities and contradictions, indicating that rearing experiences did influence the notion of their "self" and probably left long-lasting effects in their development. It is safe to say that there have been as many outcomes in terms of identity feelings as the number of identified children.

The horrific experience of these children and their families has raised important questions about the relative role and interaction of the different dimensions of personal identity, at least in the Argentine context at the edge of life and death. While it is true that each individual has a characteristic genetic makeup, "a person's identity should not be reduced to genetic characteristics, since it involves complex educational, environmental and personal factors and emotional, social, spiritual and cultural bonds with others and implies a dimension of freedom." (UNESCO 2003). In the Argentine context of disappeared children, the controversy over *nature versus nurture* in shaping personal identity took many forms and is far from settled. On the one hand, statements like *"Now I know why I became a political activist myself, as if the genes of my mother influenced my political behavior even though I did not know she had existed"* (Donda 2009) or *"I fight for the same ideas that my parents fought, because that was passed to me through their DNA"* (Sandoval-Fontana 2010) could be interpreted as if genomics would trump rearing in determining identity, and that political ideology is determined by genes. This interpretation, however, runs counter to the diversity of reactions observed among different victims, including discordance between siblings who had both been disappeared, either in the same or in a different home. In fact, those reductionistic expressions attest more to current DNA mystique in Argentina than to truly felt beliefs of those who expressed them.

It is interesting to speculate on the degree to which these extreme experiences may have influenced in Argentina the views on the relative roles of nature and nurture in human development in general, and in personal identity in particular. In fact, these theories have always been context dependent, historically, ideologically, and politically, While in the first few decades of the twentieth century, a neolamarckian view of the influence of rearing and environment in health development prevailed in academic and ideological circles permeated by Catholicism and conservative politics (Vallejo and Miranda 2005), a more reductionistic view of human nature started to take hold as developments in genomics occurred in the late twentieth century. Contrary to those polar views, the diversity of outcomes observed among the appropriated children described in this chapter, attest to the great complexity of genetic and environmental factors in shaping personal identity throughout a person's lifetime, demonstrating the futility of the *nature* and *nurture* controversy.

Conclusions

Argentine society is still coming to grips with the recent past of repression and egregious violations of human rights. The judicial system has only recently begun to tackle this issue, enforcing the duty of the state to bring to justice those responsible of forced disappearances and suppression of identity of children. The appropriated children are now in their thirties, and the 105 individuals who have recovered their true identity are essentially thrilled about it, even though this happened through a very painful process and with still-unknown long-term consequences for mental health and development. It is to be noted, however, that for each individual who recovered his/her identity there are at least two that are still victims of the crime of suppression of identity, which is a crime against humanity.

As to the role of genetics in society, it is an important sign that the United Nations Council on Human Rights recently passed a resolution submitted by Argentina that declares forensic genetics a fundamental tool to investigate violations of human rights such as suppression of identity and disappearance (United Nations 2009). The role of genetics in Argentina in bringing to light the true identities of appropriated children and the UN resolution goes a long way in redressing the abuses against human rights committed in name of genetics in the past, and reconcile the science of genetics with the well-being of humankind. It is of interest that the Argentine experience in genetic identification of children of the disappeared maintains the ambiguity of the relative roles of nature and nurture in human development and behavior. While many young men and women who recovered their true genetic identity identified with the activism of their disappeared parents and became progressive political activists themselves, others did not. Undoubtedly identity is a complex phenomenon that cannot be reduced to genetic characteristics, since it involves complex educational, environmental, and personal factors and emotional, social, spiritual, and cultural bonds with others and implies a dimension of freedom. The interdisciplinary work by social scientists and geneticists on the meaning of personal identity in the development of human beings will further contribute to healing the sequelae of the dictatorship that still affects the social fabric of Argentina.

Notes

1. For more information, see www.conadi.jus.gov.ar
2. For more information, see http://www.mincyt.gob.ar/ministerio/estructura /unid_asesoras/com_abndg/index.php

References

Abuelas de Plaza de Mayo (1995) *Filiación, Identidad, Restitución: 15 años de lucha de* Abuelas *de Plaza de Mayo (Genetics, Identity, Restitution: 15 Years of Struggle of* Abuelas *de Plaza de Mayo).* Buenos Aires, El Bloque Editorial. Available in Spanish at http://www.conadi.jus.gov.ar/home_fl.html (accessed 13 April, 2011)

———. (1997) Restitución de niños *(Restitution of Children).* Buenos Aires, Eudeba. Available in Spanish at http://www.conadi.jus.gov.ar/home_fl.html (accessed 13 April, 2011).

———. (2008a) *La historia de* Abuelas. *(The History of* Abuelas*).* Buenos Aires, Abuelas de Plaza de Mayo. Available in Spanish at http://abuelas.org.ar/material /libros/abuelas30.pdf (accessed 13 April, 2011).

———. (2008b) *Las* Abuelas *y la Genética (The* Abuelas *and Genetics).* Buenos Aires, Abuelas de Plaza de Mayo. Available in Spanish at http://abuelas.org.ar /material/libros/LibroGenetica.pdf (accessed 13 April, 2011).

———. (2008c) Centro de Atención por el Derecho a la Identidad. In: Lo Giudice (ed.) *Psicoanálisis: Identidad y Transmisión (Psychoanalysis: Identity and Transmission).* Buenos Aires, Abuelas de Plaza de Mayo. Available in Spanish at http://abuelas.org.ar/material/libros/libro_psico_08.pdf (accessed 13 April, 2011).

Amnesty International (1979) *The "Disappeared" of Argentina.* London: Amnesty International.

Anderson, M. E. (1993) *Dossier secreto: Argentina's desaparecidos and the myth of the "Dirty War."* Boulder: Westview Press.

Arditti, R. (1999) *Searching for Life: The Grandmothers of the Plaza de Mayo and the Disappeared Children of Argentina.* Berkeley: University of California Press.

Argento, A. (2008) *De vuelta a casa: Historias de hijos y nietos restituídos. (Coming Back Home: Stories of Restituted Children and Grandchildren).* Buenos Aires: Marea Editorial.

Bar, W. et al. (1997) DNA recommendations: Further report of the DNA Commission of the ISFG regarding the use of short tandem repeats systems. *International Journal of Legal Medicine,* 110, pp. 175–176.

Carracedo A. et al. (2000) DNA Commission of the International Society of Forensic Genetics: Guidelines for mitochondrial DNA typing. *Forensic International Science,* 110, pp. 79–85.

Cohen-Salama, M. (1992) *Tumbas anónimas: Informe de la identificación de víctimas de la represión ilegal. (Anonymous Graves: Report on the Identification of Victims of the Illegal Repression)* Equipo Argentino de Antropología Forense. Buenos Aires: Catálogos Editora. More available at www.eaaf.org

CONADEP (1986) *Comisión Nacional sobre la Desaparición de Personas: Nunca Más. (Never Again: Report by Argentina's National Commission on Disappeared People).* London: Faber.

Di Lonardo A. M. et al. (1984) Human genetics and human rights: Identifying the families of kidnapped children. *America Journal of Forensic Medical Pathology,* 5, pp. 339–347.

Donda, V. (2009) *Mi nombre es Victoria: Una lucha por la identidad (My name is Victoria: A struggle for identity).* Buenos Aires: Editorial Sudamericana.

EAAF Annual Reports (1990–2009) Equipo Argentino de Antropología Forense (Argentine Team of Forensic Anthropology). Available at *www.eaaf.org.* (accessed 13 April, 2011).

Elkin, M. (2008) Maria Barragán Succeeds in Getting Adoptive Parents Jailed. *The Times Online,* April 5, 2008. Available at http://www.timesonline.co.uk /tol/news/world/us_and_americas/article3687055.ece. (accessed April 13, 2011).

Gill, P. et al. (2001) DNA Commission of the International Society of Forensic Genetics: Recommendations on Forensic Analysis of y Chromosome STRs. *Forensic Science International,* 124(1), pp. 5–10.

Ginther, C., Issel-Tarver, L., and King, M. C. (1992) Identifying individuals by sequencing mitochondrial DNA from teeth. *Nature Genetics,* 2 (2), pp. 135–138.

Harvey-Blankenship, M. and Shigekane, R. (2010) Disappeared children, genetic tracing and justice. In: Parmar S. et al (eds.) *Children and Transitional Justice: Truth-Telling, Accountability and Reconciliation.* Cambridge, MA: UNICEF and Harvard University Press, pp. 293–326.

International Criminal Court (2002) Rome Statute, 2187 U.N.T.S. 90, entered into force July 1(Article 7).

Joyce, C. and Stover, E. (1991) *Witness from the Grave.* New York: Ballantine Books.

King, M. C. (1991) An application of DNA sequencing to a human rights problem. *Molecular Genetic Medicine*, 1, pp. 117–131.

Larmer, B. (1993) The lost generation: The children of the disappeared are a painful legacy of Argentina's "Dirty War." *Newsweek*, February 8.

Martinez, D. (2010) Francisco Madariaga Quintela, Nieto No.101 recuperado (Recuperated grandchild No. 101). *Página 12*, February 28.

Martínez-Marignac V. L. et al. (2004) Characterization of admixture in an urban sample from Buenos Aires, Argentina, using uniparentally and biparentally inherited genetic markers. *Human Biology*, 76, pp. 543–557.

Morling, N. et al. (2002) Paternity Testing Commission of the International Society of Forensic Genetics: Recommendations on genetic testing in paternity cases. *Forensic Science International*, 129, pp. 148–157.

Nash, N. C. (1993) Argentines contend for war orphans' hearts. *New York Times*, May 11.

Organization of American States (OAS) (1978) *Report on the Situation of Human Rights in Argentina*. Washington: Organization of American States.

Owens, K. N., Harvey-Blankenship, M., and King, M. C. (2002) Genomic sequencing in the service of human rights. *International Journal of Epidemiology*, 31(1), pp. 53–58.

Penchaszadeh, V. B. (1984) Aplicación del análisis de marcadores genéticos para la identificación de niños desaparecidos (Application of genetic marker analysis for the identification of disappeared children). Report of a Consultation to the Panamerican Health Organization. Washington. (Unpublished, available from the author).

———. (1992) Abduction of children of political dissidents in Argentina and the role of human genetics in their restitution. *Journal of Public Health Policy*, 13, pp. 291–305.

———. (1997) Genetic identification of children of the disappeared in Argentina. *Journal of the American Medical Women's Association*, 52(1), pp. 16–27.

———. (2008) Argentina: Public health genomics. *Public Health Genomics*, 12, pp. 59–65.

Sandoval-Fontana A. (2010) Quoted in: Forero j. Orphaned in Argentina's Dirty War, Man Is Torn between Two Families. *Washington Post*, February 11, 2010.

Sims, C. (1995) Argentine tells of dumping "dirty war" captives into sea. *New York Times*, March 13.

UNESCO (2003) International Declaration on Human Genetic Data. Available from www.unesco.org (accessed April 13, 2011).

United Nations (2007) *International Convention for the Protection of All Persons from Enforced Disappearance, Adopted by the un General Assembly* (article 2), December 20, United Nations.

———. (2009) *Resolution on the Use of Forensic Genetics in Cases of Serious Violations of Human Rights*. UN Council of Human Rights, A/HRC/10/L.36. United Nations.

Vales, L. (2008) Apropiadores de María Eugenia Barragán sentenciados a 8 y 9 años de cárcel. Una condena por "un vínculo cruel y perverso" (Appropriators of María Eugenia Barragán Sentenced to Eight and Seven Years in Prison: a Conviction for "a Cruel and Perverse Bond") *Pagina 12*, April 5, http://www .pagina12.com.ar/diario/principal/index-2008-04-05.html (accessed April 13, 2011).

———. (2010) "No tener identidad es como ser un fantasma:" El nieto recuperado número 101 ("Not having identity is like being a ghost:" The recuperated grandchild number 101). *Pagina 12*, February 24, http://www.pagina12.com .ar/diario/principal/index-2010-02-24.html (accessed April 13, 2010).

Vallejo, G. and Miranda, M. (2005) La eugenesia y sus espacios institucionales en Argentina (Eugenics and its institutional spaces in Argentina). In: Miranda, M. and Vallejo, G. (eds.) *Darwinismo social y eugenesia en el mundo latino* (*Social Darwinism and Eugenics in the Latin world)*. Buenos Aires: Siglo XXI Iberoamericana, pp. 145–192.

Van Boven, T. (1988) The disappearance of children in Argentina. *Report to the United Nations Commission of Human Rights*. August 10. Document E/CN.4/ Sub.2/1988/19. UNCHR: Geneva.

Contributors

Carlos Aguilar-Salinas works at the Department of Endocrinology and Metabolism at the National Institute of Medical Science and Nutrition in Mexico City, Mexico.

Sergio Avena is an anthropologist and researcher at Consejo Nacional de Investigaciones Científicas y Técnicas (CONICET). He teaches biological anthropology in the Faculty of Philosophy and Letters, National University of Buenos Aires, Argentina, and evolution at the University Maimonides. His field of study is the process of admixture in cosmopolitan populations of Argentina.

Carlos Andrés Barragán is a PhD candidate at the Science & Technology Studies Program at the University of California, Davis. His dissertation "Situating genetic expressions: Human genomic research and bio-identity in Amazonia," funded by the Pacific Rim Research Program, explores the articulations of biological capital with forms of individual and collective identity, as they emerge from human genetic and genomic research involving ethnic minorities in the Colombian and Brazilian Amazon.

Gabriel Bedoya works at the Laboratory of Molecular Genetics at the University of Antioquia, Medellin, in Colombia.

Bernardo Bertoni works in the Department of Genetics in the School of Medicine, UdelaR in Montevideo, Uruguay. He is a human geneticist interested in describing the patterns of human genome variation, admixture, or other microevolution process of a local population and how they can contribute in the understanding of complex disease. His research is focused on susceptibility genes in melanoma, breast cancer, and preterm birth.

Telma de Souza Birchal is a professor at the Department of Philosophy at the Federal University of Minas Gerais, Brazil. Other than her research and publications on modern philosophy (Montaigne, Descartes, and Pascal), she has developed research in Oxford and in Brazil on ethics and on practical ethics, and has published papers on the relation between science and ethics and also on reproductive ethics.

Desmond Campbell works in the Department of Genetics, Evolution, and Environment at University College London.

Francisco Raúl Carnese is Director of the Biological Anthropology of the Institute of Anthropological Sciences and consulting professor in the Faculty of Philosophy and Letters, National University of Buenos Aires, Argentina. He specializes in the study of human population genetics in American Indian and cosmopolitan populations of Argentina. He is currently conducting studies of genetic mixing in populations of Argentina and others countries in South America. He was president of the Association of Biological Anthropology of Argentina and the Latinoamerican Association of Biological Anthropology.

Cristina Dejean is a biochemist and teaches biological anthropology in the Faculty of Philosophy and Letters, National University of Buenos Aires, Argentina, and evolution and biochemistry at the University Maimonides. She studies the process of admixture in cosmopolitan populations of Argentina, and she is a specialist in the study of ancient DNA.

Constanza Elena Duque works at the Laboratory of Molecular Genetics at the University of Antioquia, Medellin, in Colombia.

Jose C. Florez, MD, PhD, is an assistant in medicine (Endocrine Division) at the Massachusetts General Hospital and an assisant professor at Harvard Medical School. He and his group have contributed to the performance and analysis of genome-wide association scans in type 2 diabetes and related traits. He leads the genetic research efforts of the Diabetes Prevention Program and conducts other pharmacogenetic studies. He has served as an associate editor for *Diabetologia* and is the editor-in-chief for *Current Diabetes Reports*. He is the recipient of the MGH Physician Scientist Development Award, a Doris Duke Charitable Foundation Clinical Scientist Development Award, and the Christian Pueschel Memorial Award granted by the National Down Syndrome Congress.

Liliana Franco works at the Molecular Genetics Laboratory at the University of Antioquia, Medellin, in Colombia.

Peter Fry graduated in social anthropology at Cambridge University in 1963 after completing his doctorate on religion and politics in what was then Southern Rhodesia, now Zimbabwe, at the University of London. In 1970, he moved to Brazil to help found the social science degree at the

State University of Campinas. From 1985 to 1993, he worked in the Ford Foundation in Brazil, Zimbabwe, and Mozambique. He was professor of anthropology at the Federal University of Rio de Janeiro until 2009, when he retired. Among his publications are *Spirits of Protest* (Cambridge: CUP, 1976), *Para ingles ver* (Rio de Janeiro: Zahar, 1982), *O que é homossexualidade* (São Paulo: Brasiliense, 1982) with Edward McCrae, *África no Brasil: Linguagem e sociedade* (São Paulo: Companhia da Letras, 1996) with Carlos Vogt, and *A persistência da raça* (Rio de Janeiro: Civilização Brasileira, 2005).

Natalia Gallego works at the Molecular Genetics Laboratory at the University of Antioquia, Medellin, in Colombia.

Sahra Gibbon holds a Wellcome Trust University Award and is a lecturer in the Anthropology Department at UCL. She has carried out research in the UK, Cuba, and more recently Brazil and is the author of a number of single- and joint-authored monographs and articles examining the social and cultural context of developments in genetic knoweldge and technology. This includes *Breast Cancer Genes and the Gendering of Knowledge* (Palgrave Macmillan, 2007) and a coedited collection with Carlos Novas entitled *Biosociality, Genetics and the Social Sciences; making biologies and identities* (Routledge, 2008).

Marcos Chor Maio is a senior researcher and professor of History of Science and Health at Oswaldo Cruz Foundation (Fiocruz) in Rio de Janeiro, Brazil. He has published articles on race, ethnicity, and health; Brazilian social thought; and the relationship between biomedical and social sciences. He is author of the book *Nem Rothschild, Nem Trostky: O Pensamento Anti-Semita de Gustavo Barroso* (Imago, 1992) and coeditor of *Raça, Ciência e Sociedade* (ed. Fiocruz, 1996); *Divisões Perigosas: Políticas Raciais no Brasil Contemporâneo* (Civilização Brasileira, 2007), and *Raça como Questão: História, Ciência e Identidades no Brasil* (ed. Fiocruz, 2010).

Michael Montoya is an associate professor with appointments in the Departments of Anthropology, Chicano/Latino Studies, Public Health and Nursing Science at the University of California—Irvine. His work examines the use and production of "race" in technoscience, the role of social life on human health, and the theory and creation of knowledge and power within community contexts. His recent book, *Making the Mexican Diabetic: Race, Science, and the Genetics of Inequality*, was published in 2011 by the University of California Press.

María Laura Parolin is a biologist and postdoctoral fellow of CONICET. She teaches biological anthropology in the Faculty of Philosophy and

Letters, National University of Buenos Aires, Argentina. She studies uniparental markers in Amerindian and Cosmopolitan populations of Argentina.

Maria Victoria Parra works at the Molecular Genetics Laboratory at the University of Antioquia, Medellin, in Colombia.

Sérgio D. J. Pena, MD, PhD, FRCP(C), is a medical geneticist and molecular biologist. He is currently professor of biochemistry at Universidade Federal de Minas Gerais in Belo Horizonte, Brazil and president of the Núcleo de Genética Médica (GENE). His main research interest is the study of genetic variation at the DNA level in man and the biomedical applications of this variability. Sergio Pena is a member of the Brazilian Academy of Sciences and member of the Academy of Sciences of the Developing World (TWAS). He was awarded the Great Cross of the National Order of Scientific Merit of Brazil. His publications include eight books and more than 200 medical and scientific articles in international journals.

Victor B. Penchaszadeh holds an MD degree (University of Buenos Aires), a M.Sc. degree in public health and human genetics (Johns Hopkins University), and a certificate in bioethics (Columbia University). After years of practice in pediatrics and clinical genetics, he turned to international research and teaching in human rights and public health aspects of genetics. He is professor of public health at Columbia University, head of the Argentine Center for Genetics and Public Health, president of UNESCO's Latin American Bioethics Network, and a member of the panel of experts in human genetics of the World Health Organization.

María Bárbara Postillone is biologist and PhD student of CONICET. She studies European mitochondrial haplogroups in cosmopolitan populations.

Alkes Price, PhD, is an assistant professor of Statistical Genetics at the Harvard School of Public Health and a 2010–2012 Alfred P. Sloan Research Fellow. His research interests include disease mapping in admixed populations, deconstructing the heritable components of common disease and other traits, and statistical methods for analyzing sequencing data.

Rayna Rapp is professor of anthropology at New York University where she team-teaches a course on "genes" as socionatural objects. The author of *Testing Women, Testing the Fetus: The Social Impact of Amniocentesis in America* and over sixty essays, she is now conducting collaborative fieldwork with anthropologist Faye Ginsburg on "cultural innovation in learning disabilities" examining the rapid contemporary rise in special education diagnosis for children in the United States.

David Reich, PhD, is an associate professor at the Harvard Medical School Department of Genetics. His work is unified by the theme of using genetic data to study the history of mixture in human populations. This includes recent mixture events, such as have occurred in the history of African Americans and Latinos in the last 500 years, and more ancient mixtures, such as those that occurred in India between 2,000–10,000 years ago, or between modern humans and Neandertals more than 40,000 years ago.

Laura Riba works at the Molecular Biology and Medical Genetics Centre at the Institute of Biomedicine, Universidad Nacional Autónomo de México (UNAM), in Mexico.

Maribel Rodriguez-Torres works at the Molecular Biology and Medical Genetics Centre at the Institute of Biomedicine, UNAM, in Mexico.

Andres Ruiz-Linares is professor of human genetics in the Department of Genetics, Evolution and Environment at UCL. He has carried out research on the evolutionary genetics of populations across Latin America ranging from studies on the original settlement of the continent to the analysis of the genetic history of recently established populations. His research also explores the impact of the genetic makeup of these populations for the analysis of disease susceptibility and gene mapping.

Mónica Sans works in genetics of present and past populations, population admixture, national identity, and recently, in cancer and ancestry. She held a PhD and a master's in biological sciences and a degree in anthropology and archaeology. She is currently the director of the Department of Biological Anthropology at the College of Humanities and Educational Sciences, Universidad de la República, Uruguay, and the director of the degree in human biology (Colleges of Medicine, Humanities, Dentistry, and Sciences), Universidad de la República. She is also researcher of the National Agency for Research Ministry of Education and Culture, Uruguay, and part of the program on Development in Basic Sciences. She has published more than 60 articles and 15 books or book chapters and directed or codirected around 20 projects.

Ricardo Ventura Santos is an associate professor of anthropology at the National Museum/ UFRJ and senior researcher at the National School of Public Health, Oswaldo Cruz Foundation (Fiocruz) in Rio de Janeiro,

Brazil. He has coauthored the book "The Xavante in Transition: Health, Demography and Bioanthropology in Central Brazi" (University of Michigan Press, 2002). His many coedited volumes include the books *Raça, Ciência e Sociedade* (ed. Fiocruz, 1996); *Divisões Perigosas: Políticas Raciais no Brasil Contemporâneo* (Civilização Brasileira, 2007); and *Raça como Questão: História, Ciência e Identidades no Brasil* (ed. Fiocruz, 2010).

Richa Saxena works in the Department of Medicine at Harvard Medical School in Boston and the Center for Human Genetic Research at Massachusetts General Hospital in Boston in the United States.

Guilherme Suarez-Kurtz is the Head of Pharmacology at the Brazilian National Cancer Institute and the coordinator of the Brazilian National Pharmacogenetics Network (Refargen). A pioneer of pharmacogenetic studies in the Brazilian population, his research explores the impact of genetic admixture on the conceptual development and the praxis of pharmacogenetics-genomics (PGx). He is a member of the Brazilian Academy of Sciences, senior investigator of the Brazilian National Research Council and professor of clinical and basic pharmacology at Universidade do Brasil, in Rio de Janeiro. Prof. Suarez-Kurtz did postgraduate work at Faculté de Médecine de Paris, Columbia University New York and University College London. He is the editor of *Pharmacogenomics in Admixed Populations*, published by Landes Bioscience in 2007, a collection of essays on various aspects of PGx and a database for PGx from peoples of four continents.

Marcela Tello-Ruiz works at Cold Spring Harbor Laboratory in New York, United States.

María Teresa Tusié-Luna, MD, PhD, is the principal investigator and head of the Molecular Biology and Genomic Medicine Unit from the Biomedical Research Institute UNAM and Instituto Nacional de Ciencias Medicas y Nutricion Salvador Zubiran in México City. She graduated from the School of Medicine, UNAM, and received her PhD degree from Cornell University in 1991. Her work focuses on the genetic and genomic basis of common diseases such as type 2 diabetes and various lipid disorders in the Mexican population. She holds the general coordination of the SIGMA Project in type 2 diabetes in Mexico since 2010.

Alberto Villegas works at the Molecular Biology and Medical Genetics Centre at the Institute of Biomedicine, UNAM, in Mexico.

Fuli Yu works in the Department of Medicine at Harvard Medical School and in the program of Medical and Population Genetics at the Broad Institute of Harvard and MIT in Boston in the United States.

Index